ADVENTURES IN
HUMAN
BEING

ALSO BY GAVIN FRANCIS

True North: Travels in Arctic Europe
Empire Antarctica: Ice, Silence & Emperor Penguins

ADVENTURES IN HUMAN BEING

GAVIN FRANCIS

First published in Great Britain in 2015 by
PROFILE BOOKS LTD
3 Holford Yard
Bevin Way
London
WC1X 9HD
www.profilebooks.com

Published in association with Wellcome Collection

Wellcome Collection
183 Euston Road
London NW1 2BE
www.wellcomecollection.org

1 3 5 7 9 10 8 6 4 2

Typeset in Sabon by MacGuru Ltd
info@macguru.org.uk
Printed and bound in Great Britain by
Clays, Bungay, Suffolk

A CIP catalogue record for this book is available from the British Library.

ISBN 978 1 78125 341 0
eISBN 978 1 78283 104 4

That thrice-worthy *Mercury* calls man a *great Miracle, a Creature like the Creator, the Ambassador of the Gods.* Pythagoras the Measure of all things. Plato the wonder of Wonders ... All men with one consent, call him *Microcosmos* or, The Little World. For his body is, as it were, a Magazine or Store-house of all the virtues and efficacies of all bodies, and his soul is the power and force of all living and sensible things.

Helkiah Crooke, introduction to *Microcosmographia* (1615)

for life's enthusiasts

Contents

A note on confidentiality

THIS BOOK IS A SERIES OF STORIES about the body in sickness and in health, in living and dying. Just as physicians must honour the privileged access they have to our bodies, they must honour the trust with which we share our stories. Even as long as two and a half thousand years ago that obligation was recognised: the Hippocratic Oath insists 'whatsoever in the course of practice you see or hear that ought never to be published abroad, you will not divulge'. As a doctor who is also a writer I've spent a great deal of time deliberating over that use of 'ought', considering what can and cannot be said without betraying the confidence of my patients.

The reflections that follow are grounded in my clinical experience, but the patients in them have been so disguised as to be unrecognisable – any similarities that remain are coincidental. Protecting confidences is an essential part of what I do: 'confidence' means 'with faith' – we are all patients sooner or later; we all want faith that we'll be heard, and that our privacy will be respected.

Prologue

If a man is made of earth, water, air and fire, so is this body of the earth; if man has in him a lake of blood ... the body of the earth has its ocean, which similarly rises and falls.

Leonardo da Vinci

AS A CHILD I didn't want to be a doctor, I wanted to be a geographer. Maps and atlases were a way of exploring the world through images that revealed what was hidden in the landscape, and were also of practical use. I didn't want to spend my working life in a lab or a library – I wanted to use maps to explore life and life's possibilities. I imagined that by understanding how the planet was put together I'd reach a greater appreciation of humanity's place in it, as well as a skill that might earn me a living.

As I grew older that impulse shifted from mapping the world around to the one we carry within; I traded my geographical atlas for an atlas of anatomy. The two didn't seem so different at first; branching diagrams of blue veins, red arteries and yellow nerves reminded me of the coloured rivers, A-roads and B-roads of my first atlas. There were other similarities: both books reduced the fabulous complexity of the natural world to something comprehensible – something that could be mastered.

The earliest anatomists saw a natural correlation between the human body and the planet that sustains us;

the body was even a *microcosm* – a miniature reflection of the cosmos. The structure of the body mirrored the structure of the earth; the four humours of the body mirrored the four elements of matter. There is sense to this: we are supported by a skeleton of calcium salts, chemically similar to chalk and limestone. Rivers of blood wash into the broad deltas of our hearts. The contours of the skin resemble the rolling surface of the land.

A love for geography never left me; as soon as the demands of medical training lessened I began to explore. Sometimes I found medical work as I travelled, but more often moved just to see each new place for myself – to experience variety in landscapes and peoples, and become acquainted with as much of the planet as I could. When
ravels in other books, I've tried to
the insights those landscapes have
has always brought me back to the
ıaking a living, and as the place from
nd end. Learning about the human
ırning about anything else: you *are*
tion, and working with the body has
sformational power that is unique.
ol I intended to train in emergency
tality of the night shifts, and the fleeting contact with patients, began to erode my sense of satisfaction with the work. I've taken jobs as a paediatrician, an obstetrician, and a physician on a long-stay geriatric ward. I've been a trainee surgeon in orthopaedics and neurosurgery. In the Arctic and the Antarctic I've been an expedition medic, and in Africa and India I've worked in simple community clinics. These roles have all informed the way I understand the body: emergency situations are extreme and offer a heightened awareness of human lives at their most vulnerable, but over the years some of the deepest and most rewarding insights medicine has given me have been from quieter, everyday encounters.

Latterly, I've worked from a small, inner-city clinic as a family d[illegible]

Cultu[illegible] the ways we imagine and inhabit th[illegible]s. Through my encounters with pati[illegible] how some of humanity's finest sto[illegible]ve relevance to, and resonance wi[illegible]actice. The chapters that follow lo[illegible]hose connections.

Some[illegible]ng someone with paralysis of the [illegible] just of the frustration of being una[illegible]ut of the age-old difficulty artists hav[illegible] portraying expression. When thinking about recovery from breast cancer, I've been conscious that perspectives on what constitutes healing are different for each patient. Three-thousand-year-old texts like Homer's *Iliad* can give insights about shoulder injuries, both ancient and modern, and the fairy tales we learned in the nursery eloquently explore ideas of illness, coma and transformation. The customs we bring to bear on our bodies are wonderfully diverse, something that struck me when thinking about the ways in which we dispose of the placenta and umbilical cord. Myths of struggle and redemption echo the convalescence stories going on in orthopaedic wards all over the world.

The word 'essay' comes from a root meaning 'trial' or 'attempt', and each chapter in this book is an essay attempting to explore just one part of the body, from just one of many perspectives. It hasn't been possible to be comprehensive – we are composed of a multitude of parts, and scores of ailments afflict every one of them. I've ordered the chapters from head to toe, like certain anatomy texts, though they can be read in any order. Head to toe is probably the most appropriate way to approach them – journeying alongside me the length of the human body.

Medicine has been my livelihood, but working as a physician has also delivered me a lexicon of human experience

– I'm reminded every day of the frailties and strengths in each of us; the disappointments we carry as well as the celebrations. Beginning a clinic can be like setting out on a journey through the landscape of other people's lives as well as their bodies. Often the terrain is well known to me, but there are always trails to be broken, and every day I glimpse a new panorama. The practice of medicine is not just a journey through the parts of the body and the stories of others, but an exploration of life's possibilities: an adventure in human being.

IT'S A TYPICAL MORNING in the clinic, my coffee cooling as I look down a list of thirty or forty names on a screen – my patients for the day. Many of the names I know well, but the first on the list is new to me. With a click of the mouse his medical records pop up, and at the top-left corner I notice that his date of birth was just last week. He's only a few days old; our encounter today will be the first entry into medical notes that, all being well, will follow him for the next eight or nine decades. The emptiness of the screen seems to shimmer with all the possibilities that lie ahead of him in life.

From the waiting-room doorway I call the baby's name. His mother is cradling the boy to her breast; she hears me and gets gingerly to her feet. She smiles and makes eye contact, then, with the baby in her arms, follows me back to the office.

'I'm Gavin Francis', I say as I show her where to sit, 'one of the doctors. How can I help?'

She glances down at her son, with a look of both pride and anxiety, and I watch her deciding how to begin.

BRAIN

1

N[illegible]f the Soul

> Thus strangely are our souls constructed, and by such slight ligaments are we bound to prosperity or ruin.
>
> Mary Shelley, *Frankenstein*

I WAS NINETEEN YEARS OLD when I first held a human brain. It was heavier than I had anticipated; grey, firm and laboratory-cold. Its surface was slippery and smooth, like an algae-covered stone pulled from a riverbed. I had a terror of dropping it and seeing its tight contours burst open on the tiled floor.

It was the start of my second year at medical school. The first year had been a helter-skelter of lectures, libraries, parties and epiphanies. We'd been asked to learn dictionaries of Greek and Latin terminology, strip a corpse's anatomy to the bone and master the body's biochemistry, along with the mechanics and mathematics of each organ's physiology. Each organ, that is, except for the brain. The brain was for second year.

The Neuroanatomy Teaching Laboratory was on the second floor of the Victorian medical school building in central Edinburgh. Carved into the stone lintel over the entrance was:

SURGERY

ANATOMY

PRACTICE OF PHYSIC

The greater weighting given to the word ANATOMY was a declaration that the study of the body's structure was of primary importance, and the other skills we were engaged in learning – those of surgery and medicine ('physic') – were secondary.

To get to the Neuroanatomy lab we had to climb some stairs, pass under the jawbone of a blue whale and slip between the articulated skeletons of two Asian elephants. There was something reassuring in the dusty grandeur of these artefacts, their cabinet-of-curiosities oddity, as if we were being initiated into a fraternity of Victorian collectors, codifiers and classifiers. There was a second set of stairs, then some double doors, and there they were: forty brains in buckets.

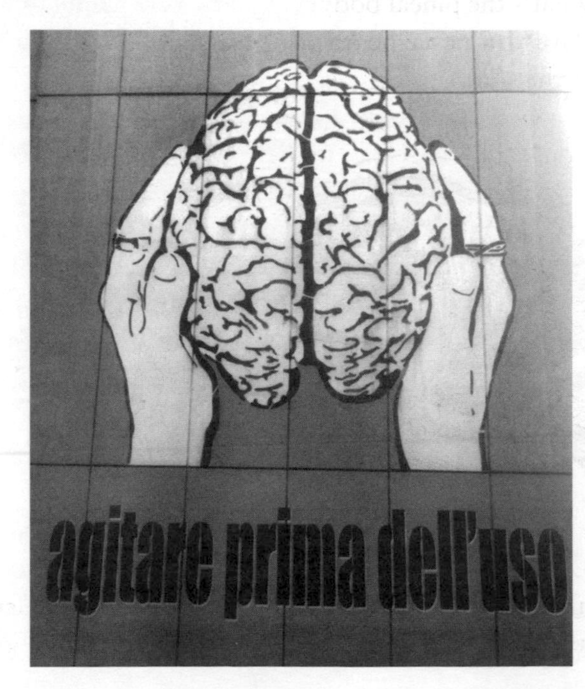

(Shake up before use)

Our lecturer, Fanney Kristmundsdottir, was Icelandic and doubled as a welfare officer, so she was also the person you were sent to see if you found yourself pregnant, or had failed an exam more than once. Standing at the front of the class she held up a half-brain, and began to point out its lobes and divisions. Seen in cross section, the brain's core was paler than the surface. Its outer surface was smooth but its interior was a complex series of chambers, nodules and fibrous bundles. The chambers, known as 'ventricles', were particularly intricate and mysterious.

I lifted a brain from its bucket, blinking at the fumes that rose from the preserving fluids. It was a beautiful object. As I cradled the brain in my hands I tried to think of the consciousness it had once supported, the emotions that had once crackled through its neurons and synapses. My dissection-mate had studied philosophy before switching to medicine. 'Hand me that,' she said, taking the brain in her hands. 'I want to find the pineal body.'

'What's the pineal body?'

'Have you never heard of Descartes? He said it's the seat of the soul.'

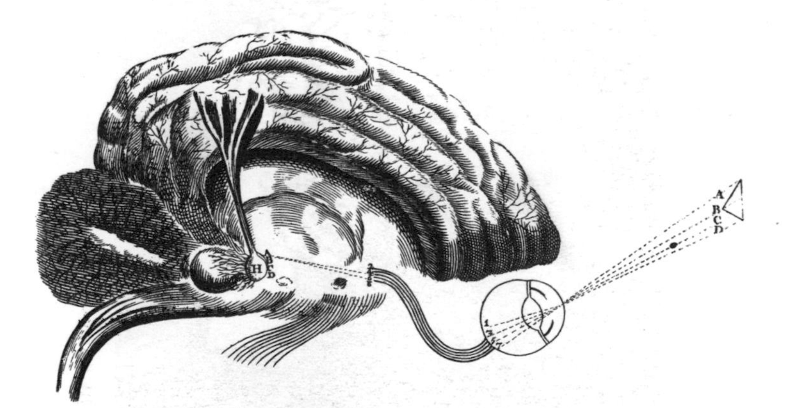

She put her thumbs between the two hemispheres, as if to open the pages of a book. At the seam that ran through the middle she pointed out a little lump, a grey pea, towards the back. 'There it is,' she said. 'The seat of the soul.'

Some years later I became a trainee neurosurgeon, and began to work with living brains every day. Each time I walked into the neurosurgical theatre I felt an urge to slip off my plastic clogs out of respect. The acoustics played a part in it: the clatter of a trolley or the whisper of an orderly seemed to echo and reverberate around the space. The room itself was a hemisphere, an upturned bowl of geodesic panels built in the 1950s. It looked the way I imagine Cold War radar domes or Dounreay's spherical nuclear reactor would appear from the inside. Its design seemed to embody that decade's belief in technology's promise of a future – an imminent future – without want or disease.

But there was still a lot of disease. I worked long days and nights with injured brains, and soon came to treat them as bruised or bloodied organs like any other. There were the victims of strokes, 'struck' dumb and paralysed by blood clots. There were creeping invasive tumours, wearing away at skulls and squeezing out personality. There were the comatose and catatonic, the car-crashed and gun-shot, the aneurysmal and haemorrhagic. There was little opportunity to think about theories of the mind, or the soul, until one day the professor – my boss – asked me to help out on a special case.

By the time I had scrubbed in and put on my gown, he was already at work. 'Come in, come in,' he said, looking up from a heap of green cloth on a table. 'You're just in time for the fun part.' I was dressed as he was; draped in the same green cloth as lay on the table, a surgical mask over my face and nose. The theatre lights flashed in the professor's spectacles. 'We're just cutting the window in the skull.' He turned back to his work, and resumed his conversation with the nurse opposite; they were discussing an American war movie. He began to cut into the skull with a saw. Smoke rose from the bone, together with a smell reminiscent of barbecued meat. The nurse sprayed water over the cutting surface, to catch the dust and keep the bone cool. She also held a suction tube to draw up the smoke, which threatened to cloud the professor's view.

Seated to one side was the anaesthetist, who wore blue pyjamas instead of a green gown; he was doing a crossword, and occasionally would reach under the pile of drapes. There were a couple of other nurses, standing back from the table, whispering to one another with their hands held behind their backs. 'Stand over there,' the professor said, and nodded to the space opposite. I jumped into position, and the nurse handed me the suction tube. I had already met the patient – let's call her Claire – and knew that she suffered from severe intractable epilepsy. Here, unusually, was someone affected not by tumour or trauma, but by a delicate shift in the electrical balance of her tissues. Her brain was structurally normal but functionally fragile, forever teetering on the edge of seizures. If normal cerebral activity – thought, speech, imagination, sensation – moves through the brain with the rhythms of music, seizures might be likened to a deafening blast of static. Claire had been so injured, frightened and handicapped by these seizures that she was prepared to risk her life with this surgery in order to be free of them.

'Suck,' the professor said. He changed the position of the tube in my hands so that it hovered over his saw blade, then began to cut through more bone. 'The neurophysiologists tell me her seizures originate just under here.' He tapped the exposed skull with a pair of forceps; the noise was like a coin dropped on porcelain. 'That's where the seizures are coming from.'

'So we'll cut out the source of the seizures?'

'Yes, but the source is very close to the area responsible for speech. She won't thank us if we make her mute in the process.'

Once he had sawn through the skull, the professor prised in little levers, similar to those used to take the tyre from a bicycle wheel, and lifted up a medallion of bone. He handed it to the nurse. 'Don't lose that,' he said. The window was about five centimetres in diameter, and revealed the *dura mater*, the protective layer that lies beneath the skull, shiny and opalescent like the inside of a mussel shell. The professor removed that too, and I looked down on a disc of creamy pink matter, ribbed like sand at low tide, with blood vessels traced over its surface in filaments of purple and red. The brain itself was slowly pulsating, rising and falling with each beat of the patient's heart.

And so to the 'fun' part, as the professor put it. The dose of anaesthetic was slowly reduced, and Claire began to groan. Her eyes flickered and then opened. The drapes had been pulled back, and the steel pins fixed into her skull were now visible.

A speech therapist had arranged her chair next to the operating table so that she was able to bend forward, close to Claire's face. The therapist explained that Claire was in an operating theatre, that she couldn't move her head, and that she would be shown a series of cards and should name each object and what could be done with it. Claire grunted, unable to nod, and they began. Her voice was drawling and disembodied – an effect of the sedatives. The cards showed images like the ones you'd find in a child's storybook. 'Clock,' she said, 'you tell the time with it.' 'Key,' she said, 'you open doors with it.' The images of simple objects went on, drawing her back to her earliest linguistic memories. Her concentration was intense, eyebrows creased, forehead glistening with sweat.

Meanwhile, the professor had swapped his saw and scalpel for a nerve stimulator. He began to dab at the surface of the brain delicately, at first holding his breath. There were no hints of bravado now, no jokes or chat: his entire attention was concentrated on two steel points separated by a couple of millimetres. The electrical effect was minimal – would barely be felt if applied to the skin – but on the sensitive surface of the brain its effect was overwhelming.

The stimulator caused an electrical storm that obliterated normal function. The portion of the brain affected was small, but it was big enough to contain millions of nerve cells and their connections.

'She carried on talking so that bit's not "eloquent",' he said. 'So we can cut it.' He placed a numbered label, like a tiny stamp, over the place he had just touched with the stimulator. The number was carefully catalogued by one of the nurses, while he moved on to the next patch. The professor called this process 'mapping': the human brain was an uncharted country being opened to surgical discovery. He moved carefully over the surface, numbering and recording; it was methodical, patient work. I had heard stories of his standing at the operating table for sixteen hours straight, reluctant to abandon the patient even to go to the toilet or eat a snack.

'Bus, you can tra … tra …'

'Speech arrest,' the therapist said, looking up at us. 'Shall we try that one again?' She showed another card. 'Knife, youah, aah …'

'There we are,' the professor said, pointing to the area he had just passed over with the electric current. 'Eloquent brain.' He placed another label carefully over the area, and moved on.

I studied the eloquent brain carefully, willing it to appear in some way different from the rest of the tissue around it. Her vocal cords and throat might make the sound, but here was the wellspring of her voice. It was the connections between neurons in that exact place, the patterns they made as they fired, that enabled speech, therefore defining it neurosurgically as 'eloquent'. But there were no distinguishing features, no sign that this patch of cortex was the channel through which Claire spoke to the world.

On one occasion at medical school a visiting neurosurgeon showed us slides of an operation to remove a brain tumour. Someone in the front row raised his hand and remarked that it didn't look like a very delicate process.

'People tend to think of brain surgeons as being very dextrous,' the neurosurgeon replied, 'but it's the plastic surgeons and microvascular surgeons who do that meticulous stuff.' He indicated the slide on the wall: a patient's brain with an aerial array of steel rods, clamps and wires. 'The rest of us just go gardening.'

Once Claire was asleep again, the professor removed a chunk of her brain – the 'epileptogenic' part – and dropped it into a bin. 'What was that chunk responsible for?' I asked him. He shrugged. 'No idea,' he said; 'we just know it's not eloquent.'

'Will she notice any change?'

'Probably not, the rest of the brain will adapt.'

THERE WAS A SCAR on her brain like a lunar crater by the time we'd finished. With her brain and mind once more anaesthetised, we cauterised the severed blood vessels, filled up the crater with fluid (so that she didn't have any air bubbles moving around inside her head afterwards), and then sutured up the *dura* with neat embroidery stitches. We reattached the disc of bone by inserting little screws through strips of titanium mesh.

'Don't drop them,' the professor said as he handed me each screw. 'They cost about fifty quid each.'

We unrolled Claire's scalp, which had been held out of the way with clips, and stapled it back in place. I met her again a couple of days later and asked her how she was feeling. 'No seizures yet,' she said. 'You could have made a nicer job of the stapling, though.' Her mouth unfurled into a triumphant smile: 'I look like Frankenstein's monster.'

2

Seizures, Sanctity & Psychiatry

Men ought to know that from nothing else but the brain come joys, delights, laughter and sports, and sorrows, griefs, despondency, and lamentations ... All these things we endure from the brain.

Hippocrates, *On the Sacred Disease*

THE PSYCHIATRIC HOSPITAL in Edinburgh looks like a stately home, set in parkland on the outskirts of the city. It was built by the city authorities as a lunatic asylum two centuries before I studied there. The idea of building an asylum had formed in the late eighteenth century – the closing years of Edinburgh's Enlightenment – as a response to the barbarity and squalor of the city centre's Bedlam madhouse.* A prominent young poet, Robert Fergusson, had died in the Bedlam in 1774, and a compassionate local doctor called Andrew Duncan had resolved to create a better institution. The new asylum was intended to be among the most compassionate and humane of its kind in Europe.

By the late twentieth century, when I arrived, the core of the original asylum had been engulfed by incongruous

* The original Bedlam, or 'Bethlehem', asylum in London gave its name to many of the lunatic asylums that were established subsequently across the British Isles.

modern architecture. There were no more lunatics (only 'patients' and 'clients') but there were laminated maps, smoking shelters, link corridors and plastic signs: 'Andrew Duncan Clinic', 'Mental Health Assessment Service', 'Rivers Centre for Post-Traumatic Stress Disorder'.

I was introduced to Dr McKenzie, the psychiatrist responsible for teaching me – a smart woman in a blue tweed jacket and skirt. She showed me around one of the in-patient wards. I was encouraged to mix with the patients, sitting with them in the smoking room and asking them how they'd come to be there. There was a wild-eyed travelling salesman with a bald pate and a silken robe: he told me he'd been admitted after unscrewing all the doors in his house because they 'blocked energy'. There was a woman who spent her time trembling in the ward's laundry cupboard and muttering to herself – she even slept there. There was a librarian, brought in by the police, who wore a waistcoat and cravat and claimed he was an incarnation of Jesus. And there was Simon Edwards, a bony, elderly man with skin like papyrus, who before being admitted to hospital had complained that his body was rotting from the inside.

Many of the patients talked incessantly, given the opportunity, but Mr Edwards did not. He spent his days sitting silently in his room, staring at the wall, immobilised by severe, psychotic depression. He wouldn't eat, didn't seem to sleep, and hardly even seemed to breathe – he gave the impression that he wanted to waste away to nothing. Dr McKenzie told me that the usual antidepressant medications had failed. As Mr Edwards was rapidly losing weight he was to begin a course of electroconvulsive therapy. If I wanted, I could come down the next morning and watch.

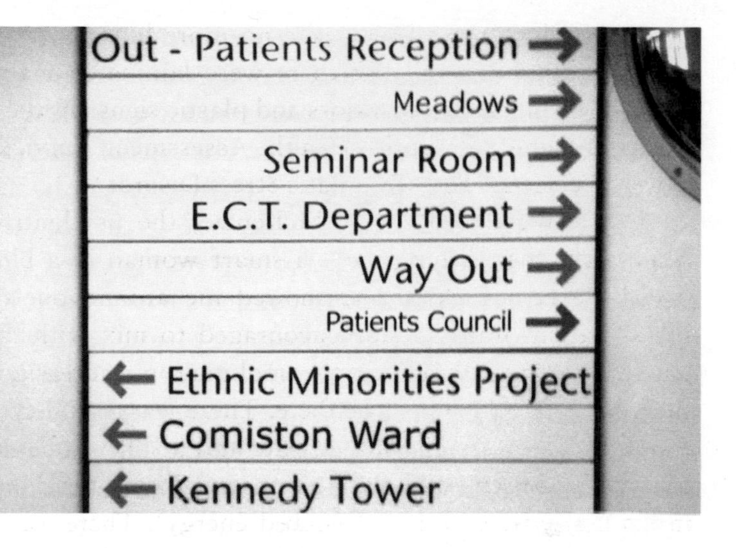

The following day I was hesitant outside the ECT department, unsure if I should enter. The door was ajar; inside I could see whitewashed walls, and a bleaching light shone in through the windows. The floor was covered in the sort of linoleum you see in operating theatres, cambered to rubber skirting boards so that dirt and germs have few places to hide. In the centre of the room was an iron-framed bed stretched with a pressed white sheet. The door swung open, pulled wide by Dr McKenzie. She had taken off her tweed jacket and was carefully rolling up the sleeves of her blouse.

There was an anaesthetist with his back to the bed; as I entered the room he turned to greet me. A medical monitor on a rolling stand stood next to the bed. There was a tray of anaesthetic drugs, a defibrillator in case of cardiac arrest and a tank of oxygen attached to a mask. All this equipment was familiar from the emergency room over at the city's main hospital, but it was startling to see it here in an environment more used to psychology, occupational therapy and pills. The ECT generator itself was a compact blue box with plugs, switches and a series of wires. It had a dashboard of ruby LEDs, like the timer on a Hollywood bomb.

Mr Edwards was wheeled in and helped onto the couch. His eyes were a coagulation of sorrow: rheumy and opaque. He said nothing, just looked blankly at the ceiling, and didn't even flinch when the anaesthetist slid a needle into his vein. He was incapable of giving consent to the ECT, and so was being treated under one of the sections of thc Mental Health Act.* Two drugs were injected into the needle: a short-acting anaesthetic and a muscle-relaxing agent, otherwise the spasm provoked by an ECT seizure can cause injury to bones and muscles. Once paralysed and anaesthetised, the patient had a plastic tube inserted into his mouth to keep his tongue from slipping back. His breathing was maintained through the mask by the anaesthetist.

Dr McKenzie placed a cylindrical metal electrode, shaped like a judge's gavel, against each of Mr Edwards's temples. She pushed a button on the handle of each and I thought I heard a low whine, like the sound of a mosquito in the ear. Mr Edwards's face quivered, his arms flexed, and his body began to twitch and shudder. 'Why is he shuddering if he's been paralysed?' I asked, wondering if something was wrong.

'These tonic-clonic movements are actually pretty minimal,' the anaesthetist said. 'If we hadn't paralysed him, they'd be far more intense.'

After only twenty or thirty seconds Mr Edwards's arms dropped to the couch. The anaesthetist rolled him onto his right side, and, after checking that all was well, pushed him on the trolley through to another room.

Dr McKenzie rolled down her sleeves and buttoned on her jacket. 'There's a lot of superstition around ECT,' she said as she reached the door, 'but it's one of the safest and, in some instances, the most effective therapies we have.'

Mr Edwards was put on a regime of two treatments a

* These 'sections' of the Mental Health Act give rise to the word 'sectioned'.

week. At first there was little improvement, but after a while his facial expression, having previously been blank, would alter when I or one of the nurses went into his room to speak to him. He seemed startled by life, like a Lazarus unconvinced that he'd been done a favour. After two weeks he began to talk.

ELECTROCONVULSIVE THERAPY is one of psychiatry's most controversial treatments – it's used less now than in decades past, but is still recommended in some cases of severe depression. It triggers epileptic seizures by applying electricity to an unconscious patient's temples – a dramatic and to some a frightening idea for a medical therapy. Seizures have long been considered an alarming transformation of the body – to the ancient Greeks they were even 'The Sacred Disease': evidence of direct communication between the human world and a spiritual realm. Fits appear to overwhelm the flesh, as if the spirit has been possessed, or has temporarily left the body. Following a seizure many people experience a period of quiet sedation, as the brain recovers to its pre-seizure state. That seizures were once considered 'sacred' is understandable – the first time I saw someone collapse with a fit, convulse, then drift off to sleep, it was as if I'd watched a process of possession, catharsis and sanctification.

Paracelsus, an alchemist-physician of the sixteenth century, called epilepsy 'the Falling Sickness'. He agreed with the ancient Greeks: epilepsy was 'a spiritual disease and not a material one', but despite its spiritual basis he insisted that seizures could respond to physical treatment, recommending a mixture of camphor (an irritant oil made from the bark of laurel trees), metal ash and 'extract of unicorn'. In the sixteenth century ingestion of camphor was known to *cause* seizures, so it was paradoxical of Paracelsus to recommend its use in epilepsy.

A major problem of the day was how to sedate lunatics

to stop them injuring themselves and others, and Paracelsus had noticed that epileptics seemed subdued following attacks. His genius was to connect the two: he wondered if by inducing seizures with camphor, he could sedate those caught up in an agitated frenzy – the first recorded instance of shock therapy. Paracelsus' influence was still felt as late as the eighteenth century: several reports were published in the 1700s describing camphor-induced seizures in the treatment of both lunacy and mania.

In the nineteenth century camphor fell out of fashion – it was too dangerous and unreliable – but the concept was resurrected in the 1930s by a Hungarian neurologist, Ladislas Meduna. Meduna had examined brains under the microscope and noticed that those of individuals who had suffered epilepsy were unusually dense with 'glia' – the supporting cells that provide scaffolding in the brain. Glial cell proliferation represents a form of scarring (the brains of boxers also exhibit this 'gliosis'). Others had reported that the brains of schizophrenics had a lower concentration of glial cells than normal, and Meduna wondered if the two observations were related. If he could induce scarring through the induction of repetitive seizures, he reasoned, maybe he could subdue madness (the same reasoning might have led him to recommend that schizophrenics take up boxing).

He began in 1934 as Paracelsus had done four centuries earlier, using camphor. But instead of using it to quieten those patients caught up in a manic frenzy, he chose subjects whose psychosis was manifest through catatonia – an unresponsive stupor. After a few camphor-induced seizures some of his patients did indeed become more responsive; he claimed a 50 per cent success rate in 'shocking' his patients back into interaction with the world. Camphor was slow and unpleasant for the patient: sometimes the seizures wouldn't come on for three hours after a painful intramuscular injection. Meduna switched to a drug called Cardiazol that works far more quickly but has horrible side effects for the patients – it initiates muscle spasms and generates strong feelings of panic. Despite this, the 1930s saw psychiatrists across Europe experimenting with Cardiazol-induced seizures in the treatment of their catatonic patients.

The 1930s were a time of reckless experimentation with the brain: the first lobotomies were performed, and a division arose between 'neurology' and 'psychiatry' that mirrored the evolving distinction between disorders of the brain and the mind. There was a feeling among those working in psychiatry that something had to be done to put it on a par with the rest of 'physical' medicine, where new treatments were being developed every year.

In 1934 a pair of Italian psychiatrists working in Rome – Ugo Cerletti and Lucio Bini – began to experiment with electricity instead of Cardiazol as a way of inducing seizures. Their first attempts involved electrocuting dogs by inserting electrodes in the mouth and anus. The dogs often died, and Bini realised that electric current traversing the heart was triggering fatal cardiac arrest. He moved to passing the current between the dogs' temples, observing that slaughterhouses in Rome used electricity in this way to stun pigs before killing them.

It took the two men a while to establish the correct voltage and current to shock a human being into a full epileptic seizure without causing death. In 1938 Mussolini was classifying political dissidents as insane, and Hitler was implementing the sterilisation of those with epilepsy, schizophrenia and alcohol dependency – Cerletti is reported to have been a subscriber to a fascist magazine. It's within this toxic political context that Cerletti and Bini selected their first patient: a man described in a later report as 'S.E.', who had been picked up babbling and hallucinating in Stazione Termini, the great railway station in Rome.

Cerletti was an academic with a high reputation and a chair at the Psychiatric Institute in Rome, but he was so anxious about the experimental nature of ECT that his trial was conducted in secret. Using equipment built together with Bini, and informed by their experiments on dogs, S.E. was restrained and a shock was administered – 80 volts of alternating current administered for just 0.25 seconds. It failed to induce a seizure, and as Cerletti prepared to increase the duration of the shock S.E. is said to have responded, 'Careful, the first was pestiferous, the second will be mortiferous!' The duration was increased a further two times – to 0.5 and 0.75 seconds, but was again unsuccessful. It was only when they upped the voltage to 110 that the shock worked, and S.E. had a full *grand mal* seizure (meaning loss of consciousness and sustained jerking of the limbs).

Reports vary. One has it that after the seizure had passed S.E. sat up with 'a vague smile' and, to the question of what had been happening to him, replied lucidly, 'I don't know, perhaps I have been asleep.' Another has him singing a popular song, and yet another that he said only 'something unemotional about dying'. But all of them agree that he became more coherent; over two subsequent months they gave him ten more shocks of what they decided to call 'electroshock therapy' (EST). On follow-up, a year later, S.E. claimed that he was 'very well', though his wife said that 'sometimes during the night he would speak as though in answer to voices'.

S.E. was the first, but there would be thousands more. As with many new treatments in medicine, doctors became its advocates before side-effect profiles and specific indications had been clearly worked out (the same was happening to lobotomy victims, who were often returned to their institutions after surgery without any follow-up). Cerletti and Bini had recommended ECT in schizophrenia, and only ten or twelve shocks, but soon treatment courses into the hundreds were being prescribed, and its indications broadened to combat the miseries of depression, anxiety, obsessive–compulsive disorder, hypochondriasis, drug addiction, alcoholism, anorexia, and conversion disorder (an extreme manifestation of psychosomatic symptoms). It was trialled in children, and to 'cure' homosexuality. In the US there were reports from state asylums where it was used as a punishment, on patients who had not finished their meal, or on those who had exhibited threatening behaviour. ECT was advised particularly in cases where patients did not have sufficient health insurance for a full course of antidepressant drugs, and to reduce labour costs on a ward. One controversial programme used repetitive ECT on a sedated patient to reduce cognitive function to the level of an infant. The aim was to 'depattern' the individual, who could start again as a 'blank slate' without psychopathology. Its author, Ewen Cameron, was later shown to have received funding from the CIA to develop 'brainwashing' techniques in which ECT would play a part.

WHY WAS IT SO DIFFICULT for Cerletti and Bini to gauge the right amount of electricity to induce a seizure? The human skull is a strong resister, comparable to the silicon used in electronics, and the threshold to elicit a *grand mal* seizure can vary as much as fivefold between different individuals because of the electrical particularities of the brain and scalp. For the first forty years of the therapy there were also great variations in the equipment used to provide the electricity: some used a mains generated sine wave of alternating current, minimally modified from the wall socket, while others provided a brief train of DC pulses. These more 'efficient' machines provoked seizures with less electricity, but psychiatrists found that they had to increase the voltage above that required purely for seizure elicitation or the therapy didn't seem to work. One of the common side effects was difficulty in speech after the seizure, thought to be caused by stunning the dominant hemisphere of the brain (in most people the left hemisphere). Attempts were made to circumvent this by applying electricity only to the right hemisphere ('unilateral ECT'), but again, the current would have to be raised above that required purely to trigger the seizure or the treatment was less effective. It appeared that the passage of electric current itself, not just the seizure, was doing something to affect the mental state of patients.

Neuroscientists can examine brain function with electroencephalograms (EEGs), which graph small changes in the electrical output of the brain by measuring at the surface of the scalp. Understanding the delicacy of neuronal function with an EEG is about as sensitive as figuring out social relationships within a city by flying over it in a bomber plane but, like aeroplane images, EEGs can convey useful information. During seizures, networks of brain cells detonate in a whirling, chaotic frenzy; the smooth meandering lines of the resting EEG switch suddenly to peaked and jagged forms, like the flames of a firestorm sweeping through the brain.

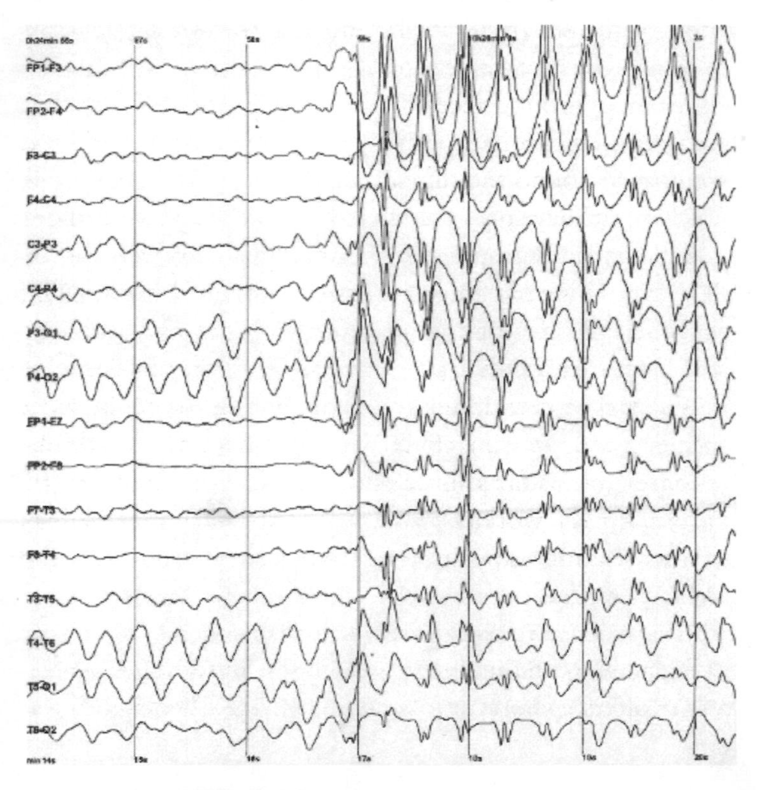

Over a usual course of ECT (in Britain and the US now just six to twelve treatments) the brainwaves between seizure treatments become slower, and the voltage and current required to trigger each seizure rises. Neurons communicate with one another across microscopic gaps called synapses, by releasing tiny quantities of chemicals called neurotransmitters. Studies in animals have shown that as a treatment course of ECT progresses, neurons become more sensitive to those neurotransmitters which dampen seizures but more resistant to neurotransmitters that encourage convulsions. It's as if the brain alters its own chemistry in an attempt to reduce the likelihood of further seizures. This alteration in brain chemistry creates poorly understood but reliably reproducible changes in mental and emotional experience.

How could changing the electrical state of the brain help those in a state of extreme mental distress? Is it the effect of the electricity itself that is of benefit, the changes in neurotransmitters caused by the seizures, or the circumstances around receiving the treatment? ECT disrupts some of the neuronal connections involved in memory, and memories around the time of treatment can be lost. Some psychiatrists have proposed that the loss of memories might even be partially responsible for the therapeutic benefits of ECT (and some patients have gone into treatment believing that the objective of the procedure was the extinction of bad memories). Other psychiatrists think that an increase in the level of certain neurotransmitters in the brain has a specific antidepressant effect. Some Freudian-minded thinkers have gone as far as proposing that the seemingly drastic nature of ECT works by offering redemption from feelings of intense guilt – a position not too distant from that of the ancient Greeks.

It's as if we're back with Paracelsus: seizures are a way of engaging with the spiritual, and invoking them with electricity offers a shortcut to a different state of being.

IT'S MORE THAN EIGHTY YEARS since Cerletti's furtive experiments with ECT, and some critics have voiced concern that the treatment is still conducted largely in secret – that it's still an arcane rite rather than a modern, medical therapy. It's as controversial as ever, though the mechanics of it are no more gruesome or distasteful than many other perfectly accepted medical or surgical procedures. For example, no one protests when electricity is used by surgeons to cauterise oozing blood vessels, but then cautery doesn't produce the unsettling transformation of a seizure.

In the last few years, psychiatrists in Scotland have tried to address the secrecy with which ECT has traditionally been carried out by setting up an open network to examine, audit and assess the experience of everyone in Scotland who receives it. The Scottish ECT Accreditation Network (SEAN) has, since 2009, published annual online anonymised reports from every hospital and clinic in Scotland that administers ECT. The psychiatrists of SEAN don't want it hidden away, stigmatised and cloaked in mystery – they've opened their work to public scrutiny in a way that other medical specialties would do well to emulate.

The reputation of the treatment in the popular imagination has been darkened by literature: in Ken Kesey's *One Flew Over The Cuckoo's Nest*, it's an instrument of torture, while for Sylvia Plath in *The Bell Jar* it's alternately terrifying and transcendental – terrifying when administered by an uncaring doctor, and transcendental when delivered by someone more compassionate. For Plath, ECT is both sacred and profane, punishment and cure – for her fictional protagonist in *The Bell Jar*, it seems to have the power of both damnation and redemption.* It's notable that in many of the deeply negative accounts of ECT in literature, the recipient wasn't sedated and anaesthetised for the treatment – modern patient experience is for most people far more benign.

* See, in particular, Sylvia Plath's poem *The Hanging Man*.

In mental health, more than in other, more 'physical' specialties, it can be difficult to define what constitutes recovery – the concept itself is slippery and whether it is achieved depends on who is asking. When Simon Edwards began to talk he described only aspects of the hospital – its food, the beds, how well he had slept. But then details about his life and his slide into despair began to emerge. 'It came over me so slowly,' he said, 'for a long time I think I didn't notice anything was wrong. It was like a heaviness on me, a suffocating fog.' Within three weeks of starting treatment, he was gaining weight. 'What changed?' I asked him, 'How do you feel different?'

'Before I could hardly move,' he said, 'I felt so weighed down. But now there's a space between me and that heaviness, a clear space.' He'd lost all memory of the days around the beginning of the treatment, and couldn't remember our first meeting. But he was no longer tormented by the belief that he was rotting from the inside; within a month of starting the treatment he was ready to go home.

On his final morning in the hospital I went to say goodbye. His wife was there, helping with his jacket, and fussing over his lapels.

'I'm fine,' he said tetchily, 'I can do it.'

'I don't know where he's been,' she said to me, 'but it's good to have him back.'

The more I began to talk about ECT with others, the more I found stories similar to that of Mr Edwards. A friend told me how helpful it had been for her grandmother; another of how her uncle's life had been saved by it. ECT is a powerful therapy – socially, psychologically and neurologically. It can cause confusion and memory loss, and disrupt the coherence of your thought. But when your habitual state is one of penetrating, paralysing misery, having the coherence of your thought disrupted may, by some, be experienced as a reprieve.

ECT is most likely to help you if your depression is in

some way 'psychotic' (you have beliefs that are manifestly untrue, such as that you are rotting from the inside) or 'retarded' (you sit silently, staring at the wall) – Mr Edwards was precisely in the group most likely to benefit. It fares less well if your misery matches one of the other headings in the evolving catalogue of despair (there are currently twenty or thirty of these, listed under F.32–F.39 in the International Classification of Diseases). Lucy Tallon, a woman who suffered recurrent bouts of depression for more than ten years, has written of how ECT was 'miraculous' in its effects – hinting at a sanctifying experience. She backs up her position by quoting Carrie Fisher, another advocate of shock therapy, for whom it 'punched the dark lights out of my depression'.

But for every positive experience of ECT that is published, there seem to be two or three that are negative; and the people with severe psychotic depression – those most likely to benefit – are perhaps the least likely to share their stories. And, as Plath attested in *The Bell Jar*, the way in which doctors speak to their patients – how compassionate, empathetic and supportive they are – can have as much influence on recovery as the physical treatment prescribed. From this perspective, increasingly recognised in psychiatric research, it's not the therapy that makes the biggest difference but the therapist.

As in many areas of psychiatry, Freud got there first: 'All physicians, yourselves included, are continually practising psychotherapy, even when you have no intention of doing so and are not aware of it.' There's nothing sacred about seizures, but there just might be something sacred about a good doctor–patient relationship.

HEAD

3

Eye: A Renaissance of Vision

Of all the things that have happened to me,
I think the least important was having been blind.

James Joyce, as quoted by J. L. Borges

MY OFFICE IN EDINBURGH has a large, east-facing window, and for most of the year I examine my patients in natural light. The exception is when a patient complains of a deterioration in their sight, and I want to look inside their eyes with an ophthalmoscope. Then it's necessary to close the blinds and feel my way in the darkness, hands outstretched, back towards the chair where the patient sits. The ophthalmoscope fires a beam of light through a small aperture, I place it close against my own eye then move within millimetres of the patient's. There are few examinations more intimate: my cheek often brushes theirs, and usually both of us, through politeness, end up holding our breath.

It's an unsettling experience, projecting an image of someone's inner eye so neatly into your own, retina examining retina through the intermediary of the lens. It can be disorientating too: gazing down the axis of the beam is like looking up into the night sky with an eyeglass. If the central retinal vein is blocked, the resultant scarlet haemorrhages are described in the textbooks as 'stormy sunset appearance'. I sometimes see pale retinal spots caused by diabetes, and they're reminiscent of cumulus clouds. In patients with high blood pressure the branching, silvered shine on the retinal arteries resembles jagged forks of lightning. The first time I looked into the curved vault of a patient's eyeball I was reminded of those medieval diagrams that showed the heavens as an upturned bowl.

Ancient Greek opinion was that vision was possible because of a divine fire within the eye – the lens was a kind of transmitter that beamed energy into the world. The flashing reflections in eyes seen by firelight seemed to confirm this theory, held by the Greek poet and philosopher Empedocles as long as two and a half thousand years ago. Part way through a series of metaphors comparing the eye with the moon and sun, he wrote: 'As when a man, about to go forth, prepares a light and kindles a blaze of flaming fire ... just so the Fire primeval once lay hid in the round pupil of the eye'.

Two centuries later Plato thought the same, though

Aristotle, who believed that light was unique in obeying the same laws whether in heaven or earth, began to question the theory – if our eyes themselves clothe the world with light, why can we not see in the dark? In the thirteenth century the English philosopher Roger Bacon hedged his bets: the soul reaches out from the lens in a projection which 'ennobles' our environment, but that environment projects itself back into the eyes.

By the seventeenth century, classical perspectives on vision were giving way. Astronomers, whose very business was the elucidation and understanding of light, were peering into the eye in order to better comprehend the stars. The astronomer-mystic Johannes Kepler was the first to write about how an image of the world was projected upside down and back to front onto the retina. When Isaac Newton was working out the motion of the planets around the sun he embarked on dramatic experiments to test the reliability of his own vision. Inserting a long blunt needle (a 'bodkin') into his own eye socket between the bone and the eyeball, he described how wiggling it around distorted his vision. Understanding didn't progress a great deal from Newton until the twentieth century, when quantum theory and the relativity theories of Einstein began again transforming our understanding of how light works.

If you are sitting reading this in the sunshine, the photons reaching your retina were born, just eight and a half minutes ago, through nuclear fusion in the core of the sun. Five minutes ago they were streaking past the orbit of Mercury, two and a half minutes ago they outran Venus. Those not intercepted by the earth will pass the orbit of Mars in about four minutes' time, and Saturn in just over an hour. After this journey across space, and in unchanging time (because, as Einstein figured out, moving at the speed of light brings time to a standstill), the sun's white light envelops the world around us and breaks into a multicoloured scatter. That scatter is funnelled by the cornea and the lens of the eye before tumbling onto the safety net of the retina. The energy of that impact causes proteins within the net to bend, starting a chain reaction, which, if enough proteins twist, leads to the firing of a single retinal nerve, and the perception of a single scintilla of light.

We can taste what's in our mouths, touch what's within our reach, smell within hundreds of metres and hear within tens of miles. But it's only through our vision that we are in communication with the sun and stars.

JORGE LUIS BORGES'S *The Book of Imaginary Beings* was first published two years after its author succumbed to the 'slow nightfall' of blindness, which he had been suffering since birth through a combination of cataracts and retinal detachments. I couldn't have looked into Borges's eyes with an ophthalmoscope: the vault of his retina was collapsing, and clouds of cataract forming in his lens would have obscured the view.

The Book of Imaginary Beings sets aside a whole page to the discussion of 'Animals in the Form of Spheres'. The greatest of these, Borges believed, was the earth itself, which was thought to be a living being by thinkers as distinguished and diverse as Plato, Giordano Bruno and Kepler himself. Borges quotes Kepler's vision of the earth as a vast orb 'whose whalelike breathing, changing with sleep and wakefulness, produces the ebb and flow of the sea', and describes the sphere as the simplest, the most beautiful, the most harmonious of forms, because every point on its surface is equidistant from its centre. The grief Borges felt at the loss of his sight surfaces fleetingly when he points out that the spherical shape of the earth recalls the human eye – 'the noblest organ in the body' – as if our eyes are themselves celestial bodies in miniature.

I was taught ophthalmology by a gifted surgeon with the exotically syncretic name of Hector Chawla. He delighted in pointing out that although ophthalmologists call the eyeball a 'globe', it is not in fact shaped like a planet, but more like a deep brandy glass.* Its stem, the optic nerve, has its foot deep in the darker recesses of the brain, while the scoop of its bowl is silvered in light-sensitive nerve fibres – the retina. In Chawla's handouts the lens, iris and cornea were like a cap placed over a glass.

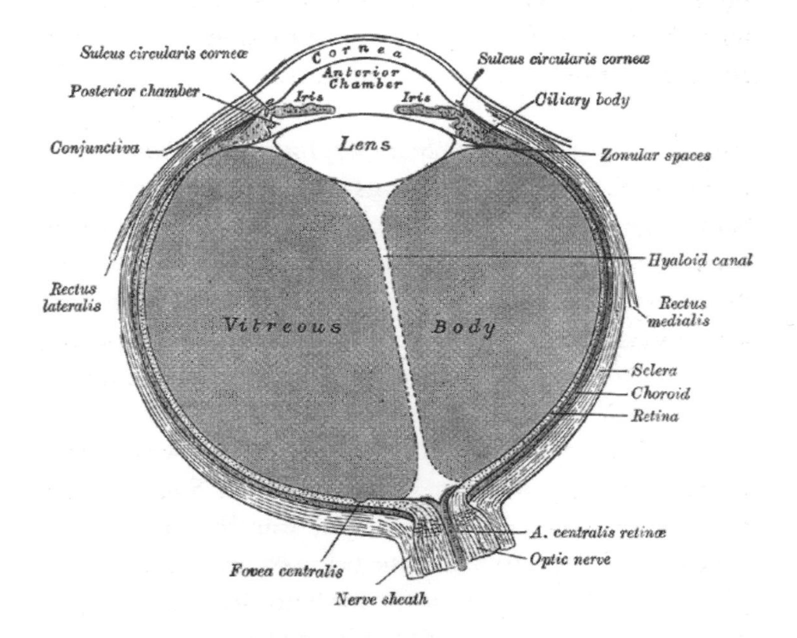

For many medical practitioners, ophthalmology seems as wrapped in mystery as alchemy, but Chawla told us how to examine the eye in clear, no-nonsense language: 'Ophthalmology tends to be thought of as a blend of mysticism and the application of drops four times a day,' he said; 'although the eye is happiest when shut, it has to open to be of any use.' Like Newton or Kepler he used astronomical metaphors to explain the functions of the eye: 'Parallel rays of light from infinity focus on the macula without effort, like a convex lens

* Very few celestial bodies are true spheres either. The earth is an 'oblate spheroid' flattened at the poles. Neither is the moon a sphere: it projects towards our planet the way the cornea projects from the eye.

which concentrates the sun into charring a piece of paper.' To assess the depth of the chamber at the front of the eye, he advised us to carry out the 'eclipse test': shining a torch onto the iris from the side to reveal its convexity, just as the curvature of the moon is revealed by the lateral rays of the sun.

BORGES INHERITED HIS WEALTH and patrician sensibility from his mother, but his love of literature, and his blindness, came from his father and paternal grandmother. Ophthalmologists struggle to agree on the cause of the Borges family's blindness, but it seems likely that glaucoma – a pathological increase in the pressure of the fluids inside the eye – was a prelude to the cataracts that overtook them.

Shakespeare, wrote Borges, was not quite accurate in describing the world of the blind as dark: his vision was obscured not by blackness, but by roiling mists of green light. He preferred Milton's greater subtlety; Milton, who ruined his eyes writing anti-monarchist pamphlets and whose 'dark world and wide' conveyed the way the blind are obliged to move tentatively, hands outstretched. Borges identified too with the way Milton composed poetry, holding it in his memory – as Borges was later obliged to – 'forty or fifty hendecasyllables' at a time, and dictating them to visitors as they called by. It was a bitter irony that the year in which Borges took up his position as the director of Argentina's National Library was also the year he lost his sight. He found himself wandering a labyrinth of a million books, but unable to read.

Photographs of Borges show him with a transcendent squint, as if one eye were watching the world, while the other bore witness to events on the astral plain. As his vision faded he lost his perception of colours at different rates. Red was the first to go, and was the most strongly mourned – his essay 'Blindness' contains a roll call of the names by which it is known in some of the tongues he knew: '*scharlach, scarlet, ecarlata, écarlate*'. Blues and greens merged into one another, and only yellow 'remained faithful' to him. The yellows of gold haunted his dreams; half a century after visiting the tiger compound at Palermo zoo he wrote a collection of poems called *The Gold of Tigers* grieving his lost sight, but elsewhere his writings suggest that he became reconciled to it. In his poem 'A Blind Man' he paraphrases Milton: 'I repeat that I have lost only / the vainest superficialities of things.'

The onset of Borges's own blindness could have unmade him, but though he experienced grief at the loss of his sight he fell rapturously into what he described as 'that literature which exceeds the life of a man, and even generations of men' – the literature of the English language. It was after he became blind that he embarked on a study of two of the roots of English: Anglo-Saxon and Old Norse. In his office at the National Library in Buenos Aires he'd gather students around him for sessions of reading the medieval classics of another continent: Beowulf; *The Battle of Maldon*; the Prose Eddas; the Volsunga Saga. 'Each word was a kind of talisman we unearthed', he wrote of the sessions with his students, 'we became almost drunk'. Just as the constellations become visible only in darkness, it was through the slow nightfall of his blindness that it became clear to him how much literature he still had to explore.

ONE OF MY TUTORS at medical school tried to encourage me into a career as an ophthalmologist. He wasn't an eye specialist himself – his field was the treatment of cancers in children. He told me that some of his patients had a survival rate of less than 50 per cent despite the best chemo and radiotherapy. He was compassionate, able, committed and enthusiastic, but when children died their parents' need to blame someone meant that he was often sued. 'Happens all the time,' he told me once in his office, while half-reading another letter of litigation. 'People get overwhelmed by grief. Now, about your career … have you ever thought of Ophthalmology?' I watched his expression as he tossed the letter to the side, exhaustion draining for a moment the colour from his face. 'Imagine how wonderful it would be,' he said, his face brightening, 'to give your patients the gift of sight!' Most ophthalmologists spend part of every week restoring sight through the removal of cataracts. 'Think how grateful they would be,' he added.

The word 'cataract' comes from the Greek *kataraktes*, meaning 'waterfall' or 'portcullis' – a barrier that descends on the vision. Cataracts develop through an opacification in the lens, and they have been treated surgically for at least two thousand years. Instructions or instruments for slicing the cornea and displacing the clouded lens away from the line of sight have been unearthed by archaeologists and historians in India, China and Greece. The removal of the lens restores only a partial, blurry kind of sight, but in the seventeenth century this displacement of the lens ('couching') had become a fairly common operation in the West. In 1722 a Frenchman named St Yves managed to fully remove a cataract, rather than just displace it deeper into the eyeball. It took only a couple of minor modifications to develop the cataract surgery that we know today.

The operation once required extraordinary self-control on the part of the patient, who had to hold his or her head and eye steady through terrible, piercing pain, as the globe of the eye was sliced open and the lens gouged out. Thanks to anaesthetic drops and paralysing agents that's no longer necessary; when I went to watch a colleague perform cataract surgery I saw the patient lie back with ease, looking up into the theatre lights as if stargazing. 'What do you see?' I asked her as she was preparing to have her eye cut open. 'Just patterns,' she said, 'just moving light and shadow. It's rather beautiful.'

After numbing the eye with drops, my colleague placed little retractors of rounded wire under her eyelids to spring them apart. Ophthalmologists have to be among the most dextrous of surgeons – shaky hands can't manage the subtle movements required to manipulate the lens. A tiny knife, shaped like a trowel only a couple of millimetres wide, sliced an entry wound into the edge of the cornea, then the space between the cornea and the lens was suffused with a synthetic jelly to maintain its pressure. Another wound was made at a separate point on the circumference of the cornea in order to introduce an instrument for manipulating the cataract, then a 'phacoemulsifier' was inserted into the first incision: it pulsed jets of fluid in and out of itself forty thousand times a second. The vibrating shock of that fluid broke up the portcullis of the cataract, simultaneously sucking the debris away. Tiny pieces of the remaining cortex were vacuumed up, and the eye left lens-less for a few moments while the surgeon prepared a replacement.

Artificial lenses can be customised according to the patient's optical prescription; they may wake up not only with their vision restored but with little need for spectacles. The lenses are of thin and flexible silicone or acrylic,* held in place behind the iris with little struts that eliminate the need for sutures. The surgeon folded the malleable new lens in half, as if rolling a calzone pizza, and inserted it through one of the incisions. Once it was in position he released his hold on the forceps and the struts sprang into position. The cataract had been removed and the lens replaced, and the whole procedure had taken just six or seven minutes. The incision was so small that no stitches were needed to close it.

FOR BORGES, sight was a transient blessing: he always knew that one day it would be withdrawn; when it was gone he turned to literature for consolation. We'll never know what revolutions in perspective he could have described for us if his sight had been restored.

I've often asked my own patients how it felt to have

* Acrylic's usefulness in the eye was discovered during the Second World War. Spitfire pilots who were shot down often ended up with acrylic shrapnel from the cockpit embedded in the eye, and surgeons noticed that it didn't cause an inflammatory reaction.

their vision restored to them by cataract surgery: 'lovely', 'marvellous', 'incredible', they often say; 'the colours are so beautiful again'. Wanting to understand more, I turned to a book on the subject by John Berger, who had his cataracts removed in 2010.

Berger has spent his life thinking about seeing. Here is his description of lying on grass looking up at a tree, from an essay published in 1960, when he was thirty-four: 'The image of the pattern of the leaves remains for a moment before it fades, imprinted on your retina, but now deep red, the colour of the darkest rhododendron. When you re-open your eyes the light is so brilliant that you have the sensation of it breaking against you in waves.' And this is from an essay published in his 1980 collection *About Looking*: 'Shelf of a field, green, within easy reach, the grass on it not yet high, papered with blue sky through which yellow has grown to make pure green, the surface colour of what the basin of the world contains.' In 1972 he had collaborated with four others – Sven Blomberg, Chris Fox, Michael Dibb and Richard Hollis – to produce a new kind of book, an extraordinary fusion of literature and visual art: *Ways of Seeing*. Berger's aim was to challenge his readers' perceptions of the images that surround us: a seminal work that redefined art criticism.

My copy of Berger's *Cataract* has William Blake's famous maxim printed on the back cover: 'If the doors of perception were cleansed every thing would appear to man as it is, infinite.'* One of the first changes the author notes after his surgery is the newness of everything, the quality of 'firstness' that has been bestowed on the world, as if all its surfaces had been dewed with light. The second is how much blue there is, even in colours such as magenta, grey and green – blue that had hitherto been deflected by the opacities

* Aldous Huxley reused the phrase in his *Doors of Perception*. His *Eyeless in Gaza* took its title from Milton's drama *Samson Agonistes*, written twenty years after Milton lost his sight.

in the lens. This blueness restores his sense of distance, 'as if the sky remembers its rendezvous with the other colours of the earth', and just as a kilometre has been elongated, so has a centimetre. As a fish is in its element immersed in water, he perceives that as human beings we are immersed in the element of light. He compares cataracts to forgetfulness, the removal of them as a kind of 'visual renaissance' that takes him back to the first colours he registered as a child. Whites strike him as purer, blacks strike him as heavier, their essential natures reborn through a baptism of light.

The words in Berger's essay are accompanied by cartoons drawn by the Turkish illustrator Selçuk Demirel. The picture to accompany the last page but one is of a couple standing side by side with their arms around each other's shoulders, watching the night sky, while the taller figure points out a star or a planet. But both the figures' heads have been drawn as eyeballs, as have the celestial bodies that hover over them – the sun and stars that generate light have metamorphosed into the organs for receiving it. Like Borges's great spheres they gaze down at the figures on earth, out into the depths of space, or even forwards into the infinite literature we all still have to explore.

ONE SPRING, I was invited by John Berger to his home in France. I had written to ask about a book he wrote in the 1960s, *A Fortunate Man – The Story of a Country Doctor*, and about his unique perspective on vision. When we met we discussed light and darkness, sightlessness and vision, and how Borges felt simultaneously liberated and imprisoned by his blindness.

He mentioned the episode, recounted in his book *Here Is Where We Meet*, in which he describes a visit to Borges's grave in Geneva. Borges was led to Geneva as an adolescent by his father, who was attracted to the city by the fame of its ophthalmologists. It was 1914, and as war overwhelmed Europe the Borges family became trapped. The young Borges grew to love Geneva and, according to a story recounted by Berger, lost his virginity there to a prostitute (he suspected his father of being another of her clients). In 1986 he returned to the city to die. His companion on that final journey was Maria Kodama, his new wife, and one of the young women who'd taken his arm and helped him to blindly navigate the labyrinth of books in the National Library of Buenos Aires.

The gravestone where Berger went to pay his respects had been chosen by Kodama. It was deeply etched with a line from the Anglo-Saxon poem *The Battle of Maldon*: '*And Ne Forhtedon Na*' – Be Not Afraid. The text curved beneath a relief taken from the Lindisfarne gravestone, of Norse warriors arriving by sea. On the reverse side there was a phrase in Old Norse from one of the couple's favourite sagas, the Volsunga Saga, which the pair had once translated together: 'He takes the sword Gram and lays it naked between them.'

Berger found the grave adorned not with flowers, but a plant in a wickerwork basket. He identified it as evergreen box: 'In the villages of the Haute-Savoie', he explains in the book, 'one dips a sprig of this plant into holy water to sprinkle blessings for the last time on the corpse of the loved one laid out on the bed.'

After paying his respects, he realised that he had no flowers or plants to leave at the graveside, so offered instead one of Borges's own flower poems: 'O endless rose, intimate, without limit / Which the Lord will finally show to my dead eyes.' Borges knew both light and darkness, blindness and vision, and that there are more ways to connect with the infinite than through sight.

4

Face: Beautiful Palsy

He sees the beauty of a human face, and searches for the cause of that beauty, which must be more beautiful.

Ralph Waldo Emerson, *Montaigne*

WHEN I WAS TAUGHT facial anatomy as a medical student most of the cadavers we dissected were those of old men with thick facial skin, stiffened by stubble. Their faces might have been tough as hide, but the muscles that lay immediately beneath that skin were fragile: delicate fronds of salmon pink laced through buttery subcutaneous fat. When trying to demonstrate the muscles that give expression to our faces I'd have to proceed with care; a slip of the scalpel and they'd be stripped off with the skin.

There were differences between individual cadavers. Though death had relaxed their expressions, the development in their facial muscles suggested something of each individual's attitude when alive. The muscles with the greatest variation were *zygomaticus major* and *minor*, the function of which is to spread the corners of our mouths to a smile. Sometimes those would be thick and well defined, implying a life filled with laughter. At other times *zygomaticus* would be shrivelled to withered little strings, suggesting years of misery. Occasionally, one side would be well developed and the other not, indicating survival following a stroke, or perhaps Bell's palsy – a paralysis of just one side of the face because of a damaged nerve.

Other muscles could give hints about each person's attitude when alive: an unusually well-developed *corrugator supercilli* hinted at a perennially angry, knitted brow – from which we get the word *supercilious*. The *levator labii superioris alaeque nasi* – an extraordinarily long name for a tiny muscle – does exactly what it says if you can fight your way through the Latin: it lifts the upper lip and wing of the nostril in a snarl. The concentric fibres of *orbicularis oculi*, arranged like Saturn's rings around each eye, are needed not just for actions such as blinking, which protects the eye surface, but when contracted more forcefully they help us squint against sunlight. They also contribute to 'crow's feet' at the angles of our eyelids. It's because of variations in the way this muscle works that some people can wink with both eyes, and others only one. *Frontalis* fibres raise the eyebrows in horror or dismay, and are the cause of the stave of lines that so often wrinkle the brow. *Orbicularis oris* puckers the lips for a kiss, while *depressor anguli oris*, beneath each corner of the mouth, pulls a pout down into a frown. Sometimes I'd find a corpse whose frowning muscles had been built up to a depressing degree.

When I later became a demonstrator of anatomy, one of my jobs was to reveal these muscles in order to help students understand the way that stroke or palsy can affect the face, as well as give a grounding to those who'd one day perform Botox injections, facelifts or facial reconstructive surgery. All in all I've probably dissected between twenty and thirty human faces, but never lost the sense of privilege it afforded. Exposing each layer of the face was a process of gradual revelation, journeying from the skin, so reminiscent of life, down to the skull, so emblematic of death. The very fragility of the facial muscles enforced a level of tenderness and respect.

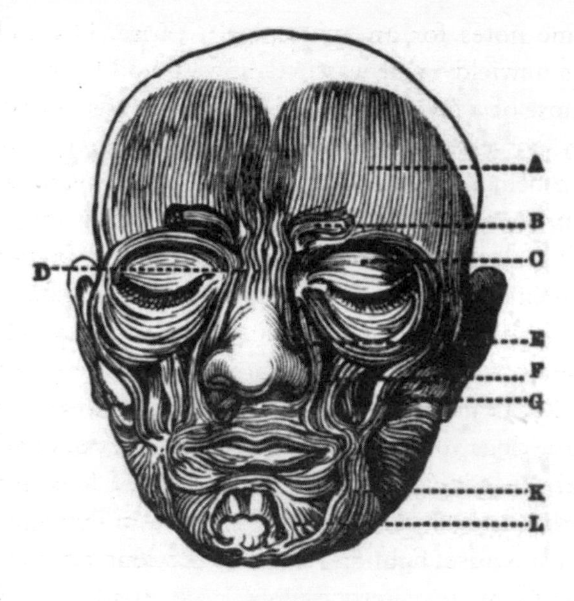

LATE IN THE FIFTEENTH CENTURY, Leonardo da Vinci, the bastard son of a Florentine lawyer, was living in Milan and thinking about facial expression more closely than perhaps anyone had before him, and few have done since. His drawings of the muscles of the face wouldn't be bettered for centuries. As a painter and draughtsman he believed in accuracy of representation, and had realised that to excel as a portraitist he had to understand those muscles intimately. He also believed that muscles were in direct communication with the soul, and that the motions of the soul could be understood through an appreciation of the body: 'The joint of the bones obeys the nerve, and the nerve the muscle, and the muscle the tendon, and the tendon the Common Sense. And the Common Sense is the seat of the soul.'

In about 1489 he was making sketches for a monumental statue of his patron's father, Francesco Sforza,* as well

* Sforza was one of Italy's most renowned *condottieri*; a kind of warlord with a private army of the sort that regularly overran Italy's city states at that time.

as some notes for an anatomical treatise. His ambition was as unwieldy as it was magnificent, and his notes offer a glimpse of a mind churning with creative and intellectual energy, possessed by a will to comprehend every aspect of human being. He intended his treatise to explain conception, pregnancy, normal and premature birth, the growth of children, the normal constitution and physiognomy of adult men and women, as well as provide a complete exposition of the veins, nerves, muscles and bones. Then he outlined how the changing expressions of the face would be the key to understanding the human condition: 'Then in four drawings you will depict the four universal conditions of man, that is: joy, with different ways of laughing, and draw the cause of the laughter; weeping in different ways, with their cause; fighting, with the different movements of killing; flight, fear, ferocity, boldness, murder ...' For da Vinci, to tabulate the actions of muscles expressing these emotions was to come close to understanding the divine source of emotions themselves. He was not interested in portraying bland representations of beauty: he wanted to capture faces as they are, as they move, whether ugly or beautiful, and if those expressions were extreme – all the better. To anatomise was to come closer to God: 'And you, man, who witnesses in this labour of mine the marvellous works of nature ... if this his composition appears to you a marvellous piece of work, you should regard this as nothing compared to the soul that dwells within that architecture.'

Later works like the *Mona Lisa* (1503–4) show just how sensitive Leonardo was to subtlety of facial expression. In the early 1490s his studio for exploring these ideas was the refectory wall of a Milanese convent on which he was painting an impression of *The Last Supper*. Other Renaissance Last Suppers had been rather staid, showing the apostles expressionlessly eating their dinner. To demonstrate the way in which emotion animates expression, da Vinci chose the moment of the Passover meal when Christ is said to have announced 'one of you will betray me'.

His apostles are caught in the commotion that followed – a drama of twelve expressions to set at play: four groups of three apostles each.* Leonardo aimed to convey a vast range of facial expression, but notable among the thirteen figures is Bartholomew, furthest to the viewer's left, who has leapt to his feet, palms on the table, and seems to glower with disbelief, furrowing his eyebrows in anger. St Andrew is third from the left: his raised palms protest his innocence and his eyebrows seem to lift in dismay.

Immediately to Jesus's left, Thomas looks bewildered, pulling the corners of his mouth down to a frown with

* Leonardo's *Cenacolo* was painted on a damp wall, and was already in a desperate state of disrepair by the mid-sixteenth century. Scholars have an idea of the power of the original from written descriptions, as well as the celebrated copy made around 1520 by Giampietrino, said by contemporaries to be the most accurate produced.

depressor anguli oris, and pointing at the ceiling with the finger which, a few days later, will be probing the resurrected Jesus's wounds in doubt. A cataclysm of emotion has broken over James the Greater who sits next to Christ; he has thrown out his arms in a fury that has darkened his eyes and creased his eyebrows to barbs.

It's said that the models for the painting were members of the contemporary Milanese elite, but it's not for its fidelity to the gospel story or for its series of commemorative portraits that the painting has been appreciated, but the way it uses facial expression as a means of giving insights into a storm of human emotion. Giorgio Vasari, biographer and contemporary of Leonardo, said that Leonardo would roam the streets, following men with particularly ugly, contorted or unusual faces in the hope of glimpsing them at moments of extreme expression. Sometimes he'd follow a particularly interesting face all the way out to the outskirts of town.

Leonardo was in Milan during a time of political turmoil; by 1499 he had to leave to escape the invading French. He trailed his sponsors to Mantua, Venice, Florence and Rome, but by the winter of 1510–11 he was back in the north, at the university and medical school of Pavia just south of Milan. Twenty years after he'd first outlined his treatise on anatomy he began in earnest the work he'd imagined so ambitiously. In the days before artificial refrigeration dissection was conducted only in winter – summer heat would putrefy corpses too quickly – and in Pavia Leonardo had a ready supply of cadavers from the hospital as well as a willing patron: Marcantonio della Torre, Pavia's professor of anatomy. Many of the anatomical sketches completed in Pavia have been lost, but from the small portion that have survived it's clear that Leonardo brought vision, imagination and astounding ability, both as an anatomist and as a draughtsman, to the task. He studied anatomy to appreciate the body as it really is rather than how it has been idealised. From his perspective the human body was the supreme culmination of God's creation.

One of the leaves from his notes shows the muscles of facial expression in precise detail, drawn more than fifteen years after he'd reproduced their effects in *The Last Supper*. *Frontalis*, which he'd shown wrinkling St Andrew's forehead, is marked as 'the muscle of fear'. He'd painted the nose and brow of Bartholomew, Peter and James the Greater snarling with rage, and here the muscle responsible for that expression, *levator labii superioris alaeque nasi*, is marked as 'the muscle of anger'. In the notes between the sketches he has written: 'Depict all causes of movement made by the skin, flesh and muscles of the face, and if these muscles have their movement from the nerves that come from the brain or not.' He perceived that there are two groups of muscles in the face: those we use for chewing, which are thick, strong and moved by the fifth nerve to issue directly from the brain, and those we use for facial expression, which are subtler, weaker and are moved by the seventh nerve from the brain.*

---

* These are the 'cranial' nerves that issue from holes in the skull or cranium, rather than 'spinal' nerves, which come out between vertebrae.

The seventh nerve is aligned to the nerve of hearing and balance; it tunnels through the skull case behind the ear and issues just under the lobe. After passing through the largest salivary gland, just behind the angle of the jaw, it splits into five branches and radiates across the face into the muscles of facial expression. The five branches are immortalised in every medical student's memory as Two Zombies Buggered My Cat (Temporal, Zygomatic, Buccal, Mandibular and Cervical). Remembering the locations of the nerve branches is helpful if someone sustains a facial injury, but is also useful in understanding the way palsy affects the ability to express emotions in the face.

I MET EMILY PARKINSON in my emergency clinic; she'd telephoned just half an hour earlier from her city-centre office. An accountant with two young children and a busy professional life, she had woken that morning to find that the left side of her face wasn't working properly. After rising she had gone into the bathroom and caught a glimpse of herself in the mirror: her left lower eyelid was sagging slightly, and when she tried to smile the left side of her face was slacker than the right. She wondered if she'd slept awkwardly on it in the night, and went downstairs to prepare breakfast. 'Look at this,' she'd said to her husband, 'half my face has gone to sleep.'

'Maybe you've trapped a nerve,' he said, and shrugged.

On the way to work she glanced in the car mirror and realised the problem hadn't gone away; in fact it was getting worse. By the time she got to work she was anxious – more so when her secretary met her with a gasp. 'What's happened to your face?' she had cried out. 'You look like you've had a stroke.'

Emily had managed to finish her make-up that morning, but a persistent tear at the corner of her left eye had smeared her mascara. On her right she had a deep fold between her nose and the corner of her mouth – the result of happily active *zygomaticus* muscles pulling on her skin for forty years – but the fold on the left was almost gone. Before, her dimples had been like parentheses, bracketing everything she said. Having a dimple on only one side gave her a partially finished, ungrammatical look.

I asked her to show me her teeth, and watched how the right side of her mouth pulled up and out, deepening her smile lines, but on the left her face hardly moved. Her wrinkles on the left had largely disappeared, but that side was lifeless. She was unable to screw up her left eye. The final test was to ask her to raise her eyebrows: the right eyebrow leapt obligingly, but on the left there was just a flicker.

The *frontalis* muscle is unusual: most muscles in the body are controlled by the opposite side of the brain, so that the right arm, for example, is moved by the left cerebral hemisphere. *Frontalis* is the exception: both sides of the brain can operate the nerve on each side. If a stroke knocks out the function of one hemisphere, victims continue to be able to raise both eyebrows, but if a nerve on one side stops working then the muscle becomes paralysed. That Emily's left *frontalis* had stopped moving meant she had not suffered a stroke.

'So, if I haven't had a stroke, what's wrong with me?' she asked.

'Bell's palsy,' I said, 'a disturbance in the nerve that gives expression to your face. It's almost certainly going to get better over the next few weeks.' I paused a moment, hoping to reassure her. 'No one is entirely sure why Bell's palsy occurs, but the nerve that controls the muscles of your face passes through a very narrow tunnel of the skull near your ear. Even slight inflammation at that site puts enough pressure on to stop it working properly.'

'What can you do about it?'

'I'm going to give you steroid tablets to take for the next ten days, to dampen any swelling around the nerve, and we should patch up your left eye to protect it.'

'What do you want to patch my eye for?'

'If the palsy progresses much more,' I told her, 'you won't be able to blink.'

THE GREEK PHILOSOPHER ANAXAGORAS, on being asked why he believed he had been born, replied 'in order to behold the heaven and the stars'. It was a common Renaissance view that humanity was special because of the way in which our faces are directed upwards.* Hairlines frame and exaggerate our naked human faces, making expressions more visible at a distance than they would have been for our hairy-faced ancestors. The whites of our eyes have expanded in comparison with other animals to make the subtlest changes in gaze and eyelid position more obvious to others. When faces are available, we pay more attention to them than to any other part of the visual world. Descriptions of the face are among some of the most lyrical and descriptive in literature; from Shakespeare's 'when forty winters shall besiege thy brow / and dig deep trenches in thy beauty's field' to Iain Sinclair's description of a character's face 'creased like a haemorrhoid cushion left too long in the bath'. Given the importance of faces to human communication, having Bell's palsy can be more than embarrassing – for some it is socially devastating.

The palsy takes its name from Charles Bell, a surgeon and anatomist of the early nineteenth century who tracked the route of the seventh nerve. Bell came from a distinguished Edinburgh family: his father had been a clergyman, two of his brothers became professors of law and another brother – John Bell – was in his day the most famous surgeon in the city. Charles hated school but loved to paint, and his mother engaged a private tutor who taught the boy to imitate the finest classical and Renaissance artists.

* Sir Thomas Browne notably pointed out that this is nonsense – the lowly flatfish has eyes directed even more piously heavenward than those of human beings.

In 1792, when Charles was eighteen, he was apprenticed to his brother John. The anatomical illustrations of their contemporaries were for the most part clumsy; Bell wrote disparagingly of the bones drawn like fence posts and muscles like rags. Charles and John worked together on illustrations for a new 'system of dissections', infusing the work with tribute to the Renaissance masters whom he had learned to imitate.

In 1809, at the height of the Napoleonic Wars, Bell was working as a surgeon and anatomical illustrator in London when the British army, with five thousand wounded men, returned to England from Corunna in Spain. He travelled down to Portsmouth to help the survivors, spending days amputating limbs, picking out shrapnel, cutting dead tissue from wounds. When he wasn't operating he was sketching, and his dispassionately accurate notebooks show figures writhing in the agonies of tetanus, gored in the abdomen, as well as gunshot wounds of the arm, chest and scrotum.

Six years later, news of the Battle of Waterloo reached London and Bell travelled to Brussels to assist. 'It is impossible to convey to you the picture of human misery continually before my eyes', he wrote from Brussels. The sketches he made this time are more detailed and involved, as if he had become more emotionally affected by the conflict; the soldiers' portraits are furnished with names and more elaborate explanations are given. Of the forty-five drawings that survive there are two notable pictures of faces; individuals whom Bell must have examined carefully for nerve damage to the face, and whose expressiveness would have suffered as a result of their wounds. One shows a soldier who had been shot through the temples by a musket ball, shattering both eye sockets and destroying the tissue behind the bridge of his nose. The other shows a man with a bullet wound in the left cheek. Without careful surgical attention both would be mortal wounds, and even despite it, the men would carry stigmata of disfigurement for the rest of their lives.

I REFERRED EMILY to the ear, nose and throat specialists, who confirmed that, for the moment, steroid tablets were the only treatment they could offer. A week later her palsy had worsened, and she was feeling more self-conscious. 'It's so embarrassing,' she said, when I called in to see how she was coping. Her fingers flickered around her face and continually brushed her hair forwards as she talked. 'I haven't been back to work, and my left eye is constantly crying. It's like I'm weeping for the loss of my face.'

Two weeks later there was no further deterioration, but neither had things improved – she was still unable to return to work. 'I couldn't bear it,' she explained, 'everyone would stare at me.' At six weeks she thought a trembling motion had come back into the corner of her mouth. 'I'm definitely dribbling less,' she told me, 'but the tears are still coming.'

'Give it time,' I said. 'Nearly everyone with Bell's palsy makes a full recovery.'

At three months that recovery seemed to have stalled,

and at six months we conceded that her palsy was unlikely to improve. She had not returned to work, and seldom went out. She had also styled her hair to hang in a curtain permanently over the left side of her face. 'I can't stand it,' she told me, 'my face scares children.'

'I'll speak to the plastic surgeons,' I said. 'They may be able to tighten up some of those slacker muscles in your affected side, and – you mentioned Botox – sometimes they use that to smooth out the good side.'

'So to treat my paralysis, they're going to paralyse my face?'

I wasn't sure they would be able to treat Emily's paralysis – it's difficult to get a damaged nerve to start working again. But in terms of normalising appearance, the most effective treatment is often to use Botox to partly paralyse the good side. 'Yes,' I said, 'I know it sounds odd, but they'll use it to make your face more symmetrical.'

BELL WAS AMBITIOUS to make his name as a surgeon – his anatomies of the nervous system were unparalleled in his day – but it was the perfection of his art that preoccupied him. Long before Waterloo, while making drawings for the *System of Dissections*, he began a prolonged study of human expression – a project similar to Leonardo da Vinci's three centuries earlier. The work was later published as *Essays on the anatomy of the expression in painting.* The book was worked and reworked throughout his life, the essays added to as Bell gained experience as a surgeon and as an artist. The final edition was enriched with his reflections from a long sabbatical in Italy, where he admired in particular Leonardo's depictions of the face. Leonardo had been forced to walk the streets looking for unusual or striking faces to paint. Bell had it easier: he had only to wait in his clinic and those faces came to him.

Thirty years after Charles Bell's death, another former Edinburgh medical student, Charles Darwin, was sufficiently inspired by Bell's work to take up the subject where he had left off. In *The Expression of the Emotions in Man and Animals* Darwin wrote '[Bell] may with justice be said not only to have laid the foundations of the subject as a branch of science, but to have built up a noble structure'. Darwin was a careful observer of the cultural as well as the natural world, and was less taken than Bell with the masterworks of Western art, particularly when it came to the study of expression. 'I had hoped to derive much aid from the great masters of painting and sculpture, who are such close observers', he wrote in his introduction, 'but, with a few exceptions, have not thus profited. The reason, no doubt, is that in works of art beauty is the chief object; and strongly contracted facial muscles destroy beauty.' He had stumbled on a paradox: we need facial muscles to express ourselves, but have traditionally idealised symmetrical, expressionless faces.

One of the few artists Darwin singled out for praise was Leonardo, for his evident belief that beauty lay also in extremes of expression, not just in neutrality. Darwin devotes a passage of *The Expression of the Emotions* to the gestures portrayed in *The Last Supper*, meditating in particular on the attitude of the apostle Andrew. One of da Vinci's maxims was that great art was arrived at through a demonstration of contrasts: 'Your painting will prove more pleasing by having the ugly set by the beautiful, the old by the young, the strong by the weak.' What would Leonardo have made of a face with Bell's palsy, when weakness and strength, ugliness and beauty, youth and age, are set side by side?

EMILY HAD HEALTH INSURANCE through her employers. The plastic surgery clinic to which I sent her was expensively carpeted, with leather couches in the waiting room and society magazines on the table. On the wall was one of the clinic's advertisements, styled as the cover of *Vogue* or

Cosmopolitan: 'Boob Jobs' and 'Tummy Tucks' took the place of the cover's splash features.

'The office was beautiful,' she laughed when she came to tell me about it. 'It was bigger than your whole building!'

The surgeon laid her on an examining couch, and cleaned the corners of her eyes, cheeks and angle of her mouth with a swab of alcohol. Then he drew up a small syringe from a vial of solution. 'He told me it would be almost painless, and it was,' Emily said, 'the needle was tiny.' He injected the solution at several points down the right side of her face, focusing on paralysing parts of her *zygomaticus* and *orbicularis oculi*, as well as Leonardo's muscles of fear and anger. 'The paralysis from these injections will be effective for four or five months,' he had said. 'And then, if you find it useful, you can come back for more.'

'So was it useful?' I asked her.

'See for yourself.' She pulled up the curtain of hair on her left side and gazed straight at me. The asymmetry was still there, but far less obvious. 'When I smile now the right side doesn't pull up and away so much' – she obliged by trying to give me a grin – 'so my face stays more neutral. It's taken years off me.'

'And do you still scare the children?'

'No, none of that,' she laughed. 'I'm delighted – even been back to work.'

AS A STUDENT AND TUTOR I'd examined the faces of the men and women I dissected with care, looking for clues about their past lives. Now, the scrutiny I'd applied to those cadavers, I turned more attentively on my patients in clinic. When I met people who'd developed frown lines too young, I began to question more about why that might be the case. I tried to distinguish those who were angry or distrustful from those who were simply afraid or feeling vulnerable, those who were anxious from those who were anguished. On meeting someone with an open, delighted face, I began asking about the secret of their happiness. And I realised that when my own expression showed irritation or impatience, relaxing my face made me feel and consult better.

In his work on facial expression, Darwin wrote: 'He who gives way to violent gestures will increase his rage; he who does not control the signs of fear will experience fear in a greater degree.' This idea, that adopting angry or fearful facial expressions, can actually *induce* feelings of anger and fear, has been borne out by psychological research. Simply by contracting Leonardo's 'muscle of anger' or his 'muscle of fear', we may become more angry or more fearful. I suspected that the reverse could well be true, that preventing expressions of fear or anger could actually diminish the experience of those emotions.

A few months later Emily came to clinic again, but this time about an injury to her knee rather than anything to do with her face. I noticed that the obviousness of her palsy was back: she must have decided against having more Botox. When I'd finished examining her knee, I asked her why.

'So you noticed,' she said, pulling back her fringe to show me her face. The deep furrow of her smile lines was back on the right, as were the crow's feet around the angle of one eye, and the furrows across one half of her brow.

'Did you get fed up of the injections?'

'Not just that, but – my feelings are more real when I can show them,' she said. 'I don't want to go through life wearing a mask.'

5

Inner Ear: Voodoo & Vertigo

For the vortex disintegrates the heavy and the light when they should be together ... Stooping causes dizziness for the same reason, for it separates the heavy and the light.

Theophrastus, *On Dizziness*

DRIVING A MOTORCYCLE is in a category apart from driving a car or even a bicycle. I'm a slow, careful rider, hesitant at speeds over sixty miles per hour, but even so there's a pleasure not just in the unaccustomed speed of motion, the ease with which a motorcycle banks into and emerges from corners, but in the blending of so much sensory information, spatial as well as visual. You become one with a motorbike in a way that's impossible in a car, and unnecessary on a bicycle.

Once I was riding a country lane on a motorcycle, late for a meeting. A forest flanked the road, its branches forming a dark canopy overhead. I wasn't so much driving as soaring through a tunnel of green, music playing through headphones in the helmet, the road unspooling ahead of me. The air felt liquid as I leaned left and right into the corners, appreciating in my sense of balance, and in the shifting weight across my muscles and joints, how my body and the bike worked the road.

Through an aperture in the trees ahead I glimpsed the stone parapet of a bridge: the road was about to take a sharp turn. I slowed the bike for the turn noting a glaze of surface green – moss on the tarmac – where the road emerged into sunlight. Abruptly the whole world shifted sideways: the back wheel had hit the moss and gone into a skid.

I was bearing down on the stone parapet at forty miles per hour, out of control. Braking hard would worsen the skid, but the stone wall was thirty yards away, then twenty, fifteen, and I slipped off the road's camber and bumped along on rubble. I was trying to keep my eye fixed on the road edge rather than on the river and its boulders below when the back wheel found purchase and, with a wobble and a swerve, I pulled up and caught the tarmac, then swerved over the bridge.

'The whole world shifted sideways' is how it felt: a momentary skid, over in a second, barely worth remarking on. But if it wasn't for the efficiency and accuracy of my sense of balance, I would have been killed.

Driving down that country road, as the back wheel of my motorcycle began its lateral slide, two events had occurred within the skull behind my ear. The bike sliding away inclined me towards the ground, tilting my head in the subtlest of angular rotations – a movement picked up by the spin of fluid through the semicircular canals within my inner ears. At the same time the jerk sideways was sensed by a related part at the base of the canal, the 'utricle', where sensitive hair cells, wired to the brain, are embedded in a jelly studded with particles of chalky material. The chalk gives the jelly mass and inertia, so as my skull made a skidding acceleration sideways, the jelly tugged on the hairs. The utricle transmits acceleration in the horizontal plane: sideways, or forward/backwards. Another part of the inner ear, the 'saccule', senses acceleration in the vertical plane.*

---

* Since 2010 many smartphones have incorporated a gyroscope and an

Just as a mammal's need for amniotic fluid in the womb is an echo of a time when all beings gave birth in the sea, the fluids within the inner ear are a reminder that once, our ancestors' balance organs were simply tubes open to seawater.* As they rolled and pitched through the three dimensions, the free flow of seawater through those tubes conveyed our motion to the brain. Though it's excluded from the usual roll call of five, balance is one of our most ancient senses: a portable sea anchor that moors us in the world.

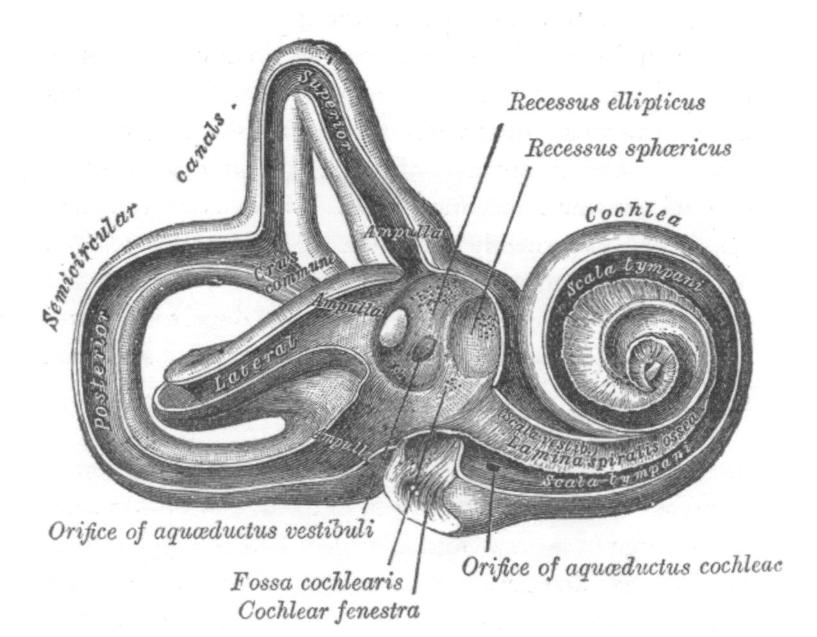

accelerometer, built with nanotechnology. Modelled on the inner ear, they orientate our phones in space.

* Some fish can't generate their own chalky material for this purpose, but as their inner ears are still open to the sea they use pieces of sand that drift in from the outside.

The word 'vertigo' is often used to describe a fear of heights, but to doctors, vertigo is the sensation of nauseating dizziness that occurs when your balance organs and your eyes give conflicting messages about your state of motion. It's related to seasickness, another result of conflicting sensory information. When you're deep in the hull of a boat in a storm, your inner ear says you're moving, but your eyes maintain that you're not. The sensation of vertigo can be just as nauseating, caused either by a diseased inner ear insisting you're stock-still when your eyes testify otherwise, or the obverse: your eyes insisting you're still while the inner ear tells your brain that you're turning.

Of all the miseries our bodies inflict on us nausea can be the hardest to bear, and one of the most difficult to treat with drugs. As a sensation it arises in a very primitive part of the brain, close to the spinal cord, suggesting that it could be a very ancient way of alerting the body to toxicity. That vertigo causes nausea probably means the brain interprets balance dysfunction as poisoning. It can be caused by infections in the inner ear, by tumours, even by washing the eardrum with warm water. It makes us retch in order to be rid of a poison, but vertigo and seasickness can't be vomited up.

JOHN WIRVELL was in his late fifties. He had a grey moustache like old rabbit fur, nicotine-stained, and his forehead was pleated by worry. Gold and silver hairs straggled from his eyebrows and gave him a startled expression. From his notes I saw that he was a taxi driver, divorcee, father of two grown-up children, and an intermittent heavy drinker. We'd met only once and he struck me as a bit of a stoic, a proud and independent man who treated doctors with caution. 'No offence,' he'd told me in the consulting room, 'but I don't really go to doctors.'

'Glad to hear it,' I'd said. 'If there's nothing wrong, why should you?'

So it was unexpected when a year or so later he requested

a visit at home, because, so the receptionist said, he'd been struck by attacks of nausea and vertigo. The attacks were so bad he was frightened to leave the house. I wondered if he'd had a stroke and called him before visiting to see if I should send an ambulance. 'Arms and legs are still working, Doctor,' he said to me down the phone. 'I just can't turn my head.'

When I arrived he was lying on his sofa, perfectly still. 'A hundred times a day the room spins, I feel like puking my guts up, and I can hardly move,' he said. 'It's been hitting me for a couple of days – when it comes I just have to lie here, praying for it to pass.'

I squatted down beside him. 'What brings it on?'

'It can be anything. Sometimes just turning to look over my shoulder sets it off. Sometimes all I need to do is roll over in bed. Bending over can do it.'

Episodes of low blood pressure can sometimes cause dizziness, but Wirvell's was slightly high. Alcohol can cause vertigo, but he'd been laying off the drink. I asked about other triggers, but he'd had no head injuries, recent infections, and had started no new medications.

'Is it always when you turn in the same direction?' I asked him.

'Yes,' he looked up at me. 'It gets worse if I look down, and to the right.'

When vertigo comes on only in certain positions it's defined, helpfully, as 'positional'. When it arrives in sudden and overwhelming bouts it's called 'paroxysmal'. The final distinction an ear specialist wants to make is between disease due to something malignant and progressive or disease caused by something benign and ultimately self-limiting. John's illness was almost certainly the latter, so he had, in the mealy-mouthed but faultlessly descriptive jargon of otolaryngology, 'benign paroxysmal positional vertigo,' or BPPV. Though the syndrome is an ancient one,* it wasn't described

* Hippocrates said that it was the fault of a southerly wind: *Aphorisms 3:17*.

until 1921 when a Viennese physician called Robert Bárány finally defined 'episodic vertigo' as a syndrome.

It used to be thought that in BPPV the chalky grains of the utricle and saccule became attached to the wrong membrane: the 'cupula' that stretches across the base of the balance canals. The grains themselves were believed to distort the cupula's shape, sending confusing messages to the brain about the head's direction of movement. Treatment focused on repetition of the nausea-triggering movements until the patient became numbed to them, which could sometimes work. In severe, recurrent cases, the skull would be opened and part of the nerve that leads to the inner ear would be severed, risking deafness. It sounds drastic, but patients affected by recurrent waves of nausea and disorientation were often thankful for it.

In the 1980s another theory was proposed, formulated by an American otolaryngologist called John Epley.* Epley believed that BPPV was caused not by chalky particles adhering to the wrong membranes, but by particles breaking free and rolling around the semicircular canals, stirring up eddies which were perceived by the brain as movement. Fashioning a model inner ear out of bits of hosepipe in his garage, he rolled it through different sequences hoping to find a way to dislodge the particles and guide them out of the canals into a less sensitive part of the organ. Using this simple technology he worked out a series of simple movements that could be performed on his office couch. When he started experimenting on real patients he found that he could cure even those who had been suffering BPPV for years. When the sequence didn't work he tried holding a vibrator to the skull behind the patient's ears just before the manoeuvre, to help dislodge any adherent chalky grains, and found that this further improved his cure rates.

---

* In the 1960s, Epley had been involved in trialling the very first cochlear implants.

Surgeons whose livelihoods were bound up with recommending expensive procedures for BPPV were sceptical, and the fact that Epley had been holding vibrators to patients' heads allowed them to label him a crank. He was laughed at in conferences, accused by some as unfit to practise. His manoeuvres were perfected in the early 1980s, but it took a decade until his harmless, effective, medication- and surgery-free treatment for positional vertigo was published in a journal respected by his peers. It took another few years to percolate into general medical clinics around the globe.

Anyone can do an Epley manoeuvre; you can download the sequence from the Internet and try it at home, even on yourself, though people with neck problems or poor circulation should be cautious. It was more than a decade after Epley published his findings that I first heard of the sequence and gave it a try. Epley reported a 90 per cent cure rate in his Oregon clinic: the results could be just as astonishing when I began to use it in Scotland.

I LED WIRVELL through to his bedroom, and asked him to sit towards the foot of his bed the wrong way round, legs extended towards the pillows. I noticed that he had small, oddly intricate ears, as convoluted as nautilus shells. I put one hand on each of them and then dropped him straight back so that his head fell past the horizontal off the end of the bed, his chin turned towards his left shoulder. This is a position that orientates the head in a certain way with respect to gravity, calculated by Epley to allow chalky grains on the left side to begin to drift down through the semicircular canals. We waited a few seconds.

'Nothing's happening,' he said, creasing his forehead. 'Is this supposed to make the dizziness go away?'

The next time I dropped him back, turning his chin towards his right shoulder, his whole body tensed and his eyes began to jerk like beads of light on an oscilloscope

– attempts by his eyes to follow the illusory motion sensed in the labyrinth. 'That's it!' he muttered, gritting his teeth, 'You're making it worse!'

In the 1950s it had been figured out that when the canals on the right side were affected by BPPV, lying back with the chin towards the right was the movement most likely to bring on an attack. After thirty seconds holding that position, the jerking of his eyes began to settle. I turned his head slowly through ninety degrees, still hanging off the end of the bed, so that now his chin pointed towards his left shoulder. His vertigo came on again, but this time less so. After another thirty seconds, I rolled him onto his left side while maintaining his chin's position, so that the position of his neck now turned his gaze to the carpet. His body relaxed, he ungritted his teeth – his symptoms were already settling. Thirty more seconds and I sat him up, asking him to slowly raise his chin and look towards the headboard of the bed.

'How do you feel now?' I asked him.

He paused a moment, then tentatively turned to look over his right shoulder. 'OK so far,' he said, swinging his legs off the side of the bed.

'Try bending over.'

He stood up, then bent his head and looked over his right shoulder – the movement that had previously set off his vertigo. 'It's like magic ... voodoo medicine!'

WHY DID A TREATMENT so simple, risk-free and effective take ten years to be reported in the medical press? It's wrong to assume that doctors are rationalists, that the medical gaze is as emptied of bias and as open to new ideas as the best science aspires to be. Physicians are just as prone to prejudice and protectionism as professionals in any other sphere of life – it's just that we rightly hold them to higher standards.

The simplicity and effectiveness of Epley's manoeuvre is like a conjuring trick, but it's also a reminder that for all the advances of modern medicine, the body and its ways can still surprise us. Physicians had been stumped for millennia about how to treat episodes of severe and incapacitating vertigo. It's heartening that it wasn't a major technical development that cracked the problem of BPPV – some new type of scanner or microsurgical procedure – but just some creative thinking, a garage, and some lengths of plastic hosepipe.

CHEST

6

Lung: The Breath of Life

On one side heavenly fire: light, thin, in every direction the same as itself ... The opposite is dark night; a compact and heavy body.

Parmenides, *On Nature*

IN ONE OF THE EMERGENCY DEPARTMENTS where I used to work there was a concealed door that led to a small yard out back. Ambulances would bring patients there when they were already dead. Rather than arriving with flashing blue lights at the main entrance there would be a discreet knock on the door, and one of us doctors would go out and certify the body so that it could be taken to the mortuary.

There are only three tasks to remember when certifying the dead: shining a torch in the eyes to see if the pupils narrow in response to the light, checking the carotid artery at the neck to feel if there's a pulse, and putting a stethoscope on the chest to hear if there's any breath. The breath is the most telling; in Renaissance times a feather would be placed on the lips to see if air was moving in and out of the lungs. The textbooks recommend a full minute of listening, but I've often done it for longer – afraid that I'd miss an agonal gasp, or one final, feeble beat of the heart. But a glance at the milky, desiccating surface of the eyes is usually enough to convince me that the dead really are dead. The yawning emptiness of the pupils is another giveaway – a glimpse into the abyss.

One night a man was brought in dead after he'd jumped from one of Edinburgh's many bridges onto the road below. His medical notes, when they arrived from the records department, said the psychiatrists had seen him earlier that week, and he'd seemed in 'good spirits'. Bystanders said he'd had no hesitation; just leapt over the parapet to his death as if he'd dropped something precious and wanted to get it back.

His was a messy corpse. His neck was badly broken and distorted, his tongue and neck were swollen, but there was little bleeding from his grazes – his heart would have stopped beating almost immediately on impact. I shone a torch into his eyes and watched the light drop into their vacant stare – there was no narrowing of the pupils, or reflection of the light from their surface. Moving onto his carotid pulse I felt something unexpected: beneath my fingertips there was a popping and crackling sensation. After checking he had no pulse I placed my stethoscope against his chest wall and heard the same crackle amplified through the earpieces. His lungs must have burst, I realised – exploded with the pressure as he hit the road. The popping and crackling was caused by air, ordinarily contained within the lungs, but now tracking out into the other tissues of the body.

Liquid and air must keep to their separate compartments in the body, just as a horizon separates the sea from the sky. Even if his eyes and lack of pulse or breath sounds hadn't convinced me he was dead, this would have. As I listened for a breath sound that didn't come, I imagined how it must feel to launch oneself from a bridge; how light and free it might feel if only gravity, and the blackness of despair, weren't pulling you down to the earth.

LUNGS ARE THE LEAST DENSE organs in the body, because they are composed almost entirely of air. The word 'lung' comes from a Germanic root *lungen*, which itself arises from another Indo-European word meaning 'light'.

Traditional Chinese, Ayurvedic and Greek medicine

all maintained that air carried invisible spirits or energies (which they called, respectively, *qi*, *prana* or *pneuma*). From those perspectives our bodies are bathed in spirit, our lungs the interface between the spiritual and the physical world. For the Greeks, as commemorated by St John's Gospel, the first principle was *logos* – the word – existence was conjured into being through sounds produced by the breath. Written texts, even those never intended to be read aloud, are often punctuated according to the needs of a speaker to take a breath.

The lungs are light as spirit because their tissue is so thin and delicate. The membranes within them are arranged so as to maximise exposure to breath, much as the leaves on deciduous trees maximise exposure to air. Just as leaves draw in carbon dioxide and leak oxygen, lungs draw in oxygen and leak carbon dioxide. If you were to stretch flat all the membranes of an adult's lungs they would occupy over a thousand square feet; equivalent to the leaf coverage of a fifteen- to twenty-year-old oak. Listening with a stethoscope you can hear the flow of air across those membranes, like the rustle of leaves in a light breeze. When doctors listen to the breath, that's what they want to hear: an openness connecting breath to the sky – lightness and the free motion of air.

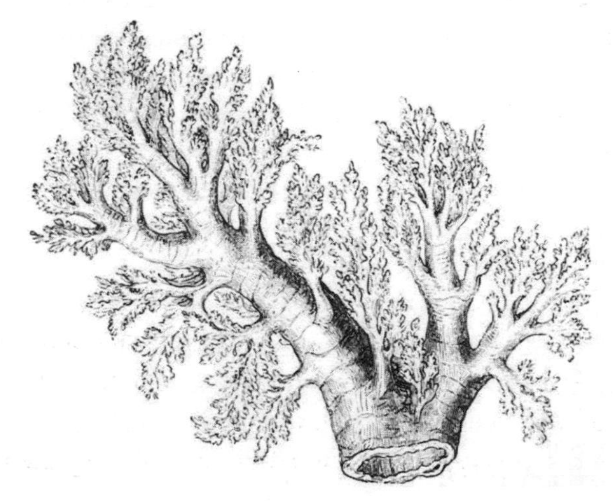

Doctors use stethoscopes to sound out solidity in the lungs: if tumour or infection consolidates the tissues, instead of the muted sigh of the breath you can hear the whistle and clatter of disease. With a stethoscope we listen for 'increased vocal resonance': the crisp transmission of words spoken by the patient. We listen for 'bronchial breathing': the sound of air whistling through the large airways. These sounds are inaudible through healthy tissue, but can be revealed by the transformed acoustics of a heavy, solidified lung. Infection rather than tumour gives rise to a third sound, called 'crepitation', when pus and mucus make the finer membranes stick to one another. Thousands of tiny air chambers then pop open and closed with each waft of breath, sounding as if the lungs had been enveloped in a thin film of bubble wrap.

When I think about lungs, the associations that come to mind are of light, airiness and vitality. When they become diseased they lose their lightness; they become ballast that pulls us towards the grave.

IT WAS A COUGH that Bill Dewart complained of first; an empty, futile cough that punctuated his sentences by day, and earned him digs in the ribs from his wife at night. Bill wore a flat cap and carried a stick, but at seventy-six was strong and still working as a plumber. He had the face of a younger man, with a surprised expression, as if startled by how stealthily age had sneaked up on him. 'What would I stop working for?' he asked me when I brought up the question of his retirement. 'Sitting around at home all day, getting under my wife's feet.'

'How much do you smoke?' I asked him, noticing the tar stains on the fingers of his right hand.

'Forty a day for the last sixty-five years,' he said, 'and I'm not about to stop now!' He laughed, the folds of his skin deepening across his cheeks: 'Cigarettes!' he said, wagging the yellowed finger at me. 'That's all you doctors want to talk about!'

I asked him to blow through a flow meter, to see how fast he could expel air from his chest. It was slower than it should have been at his age, but the smoking would explain that. After helping him unbutton his shirt, I placed my left hand against the back of his chest, and began to tap that hand's fingers with the middle finger of my right. Over healthy lungs this tapping makes a sound like a muffled drum; resonant and soft, with a slight give felt in the left hand. When lung tissue is solid or filled with fluid it's like tapping the drum's rim rather than its skin: dull and hard, without any give.

I sounded all the regions of his chest: front and back, upper, middle and lower. All sounded hollow. I tracked the same route using the stethoscope: the sound throughout was of soft rustling leaves – there was no sense of a solid portion of lung beneath my hands. Lastly I had him say 'ninety-nine' while I listened over the same areas (the 'n' sound resonates particularly well through the chest). Left and right, upper and lower, front and back, the transmitted words were soft and indistinct. There was none of the crispness that I'd expect of sound transmitted through consolidated lung.

'I don't think you've got a chest infection,' I told him. 'And I can't see that any of your tablets cause a cough.' I glanced from his face to his tar-stained fingers. 'But I'd like to do some blood tests, and send you for a chest X-ray.'

ON THE RESPIRATORY MEDICINE wards I had two teachers. One of them claimed membership of a lofty tradition in clinical examination, and instructed us to practise percussion of the chest by placing a coin beneath a telephone directory. 'Shut your eyes and tap the telephone directory,' she said, 'the acoustics over the coin are slightly different.' She insisted that examination of the lungs was a fine and subtle art that could be improved upon throughout a clinical career. The other teacher thought that this was the auditory equivalent of examining bumps in the skull to divine personality, or tasting urine for sugar. In the first tutorial he gave, he held up a chest X-ray against the window. 'This,' he said, 'is how you examine the chest. With *X-rays*.'

Bill Dewart's X-ray looked pretty normal to me. His windpipe was straight, ending in the Y-fork that branched out into each lung. His lungs themselves looked dark, without a trace of the density that can suggest tumour or infection. If anything they were *too* dark, suggesting emphysema caused by those sixty-five years of cigarettes. His heart was of normal size with respect to the diameter of his chest, and the outline of his diaphragm was distinct rather than hazy. Apart from the emphysema the only abnormality I could detect were some knuckle-like thickenings on the ribs of the right-hand side. 'Did you ever break your ribs?' I asked him.

'Aye,' he said, wincing at the memory. 'But the other guy came off worse.'

'So all in all, I can't see a reason for your cough,' I said.

'Maybe it's getting a wee bit better,' he said, but I was unconvinced.

'Let's try an inhaler, send a sample of your phlegm to the lab, and meet again in a week.'

'THE COUGH'S WORSE than ever,' he said when he came back to me. 'Not just that but my wife says I'm losing weight. I eat like a horse but I can't put anything on.' Again I tapped the different lobes of his lungs; again I listened to them, but didn't hear anything unusual. 'And that inhaler is a waste of time.'

'It's early yet,' I said. 'It's worth persevering.'

'Well don't persevere too long,' he said, 'or there won't be anything left of me.'

I started him on high-calorie drinks, and issued him with a dietician's advice sheet advising chocolate bars between meals, and smothering all his meals in cheese. I also arranged a follow-up X-ray, and wrote to the respiratory specialists asking if they'd do a CT scan of his chest.

The second X-ray report was sent electronically only a day later – the radiologist had felt it too urgent to wait for the mail. 'Comparison is made with the previous film,' it said. 'There is a degree of mediastinal widening and some deformation of the right main bronchus suggesting subcarinal lymphadenopathy. Further CT examination is recommended.'

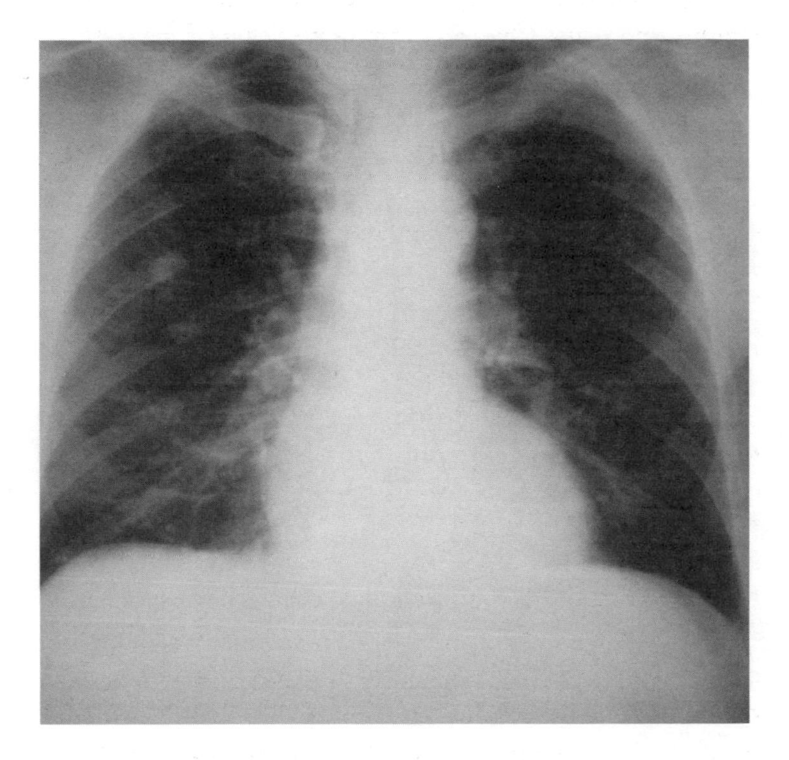

The 'carina' the radiologist was referring to was the point where the trachea splits into two separate tubes, one for each lung. 'Carina' is Latin for 'keel', and is used to describe parts of the body where two sloping planes meet a central ridge, just as the two halves of a hull meet along the keel of a boat. There are two other carinas in the body: one under an arching band of tissue in the brain, where it links parts of the two hemispheres related to memory, and one in the lower vagina where the urethra indents the vaginal wall.

The carina is the most sensitive stretch of the human airway: it's the place where any object falling down the windpipe, such as a flicked peanut or a lump of choked food, is likely to strike first. It has to be sensitive because anything falling into the lung must be coughed out immediately, otherwise infection or suffocation might follow. Swelling around it can bring about a particularly persistent and distressing cough, as the body tries to expel whatever is causing the irritation. The radiologist was suggesting that the lymph nodes under that keel of tissue had become heavy and swollen and, like a boat burdened with too much ballast, the hull of the airway was bent out of shape.

The CT scan confirmed the swollen lymph nodes around the end of Bill's trachea, as well as in the area where the airways, arteries and veins enter and leave each lung. The lymph nodes of that region drain fluid from the lung tissues, and the fact that they were swollen suggested they were weighed down with tumour cells. But there were other possibilities: infections and some unusual immune conditions. To find out which, Bill would have to undergo a biopsy.

IF YOU BLOW AIR onto your hand with your mouth open, your breath feels warm and moist. If you purse your lips and blow again your breath this time will feel cold. It's a Renaissance belief that your soul is attached most firmly to your body at the lips; after all, it's the place where the breath of life enters and leaves your body. That the breath can change between warm and cold just by altering the position of your mouth was once powerful proof of its vitality. The truth is a little more prosaic: the pursing of your lips puts the air under pressure; it's the re-expansion of that pressurised air that draws in heat from your hand and makes it cool.

When you breathe in through your nose, air is channelled

by folds in the nasal bones called 'turbinates' that roll the air like turbine blades. They slow, warm and moisten the air as it makes its way to the back of the nose, where the front of the spine makes its joint with the skull. From that angle – the 'post-nasal space' – it is redirected down behind the tongue, into the laryngeal cartilages and between the false and true vocal cords. The anatomical landscape that makes voice from breath is intricately named: *triticeal*, *corniculate* and *arytenoid* cartilages; *cuneiform* tubercle and *aryepiglottic* fold.

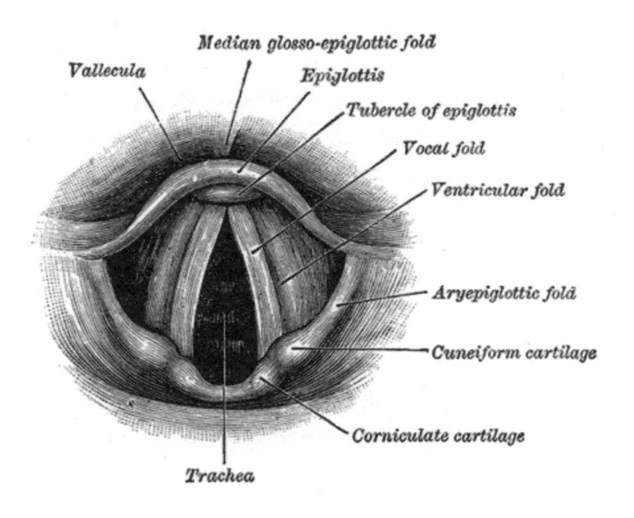

Muscles in the larynx alter the tension between these different elements giving pitch to the voice, whether we're screaming in alarm or singing an aria. From the vocal cords the breath flows on another five or six inches to the carina, then, like water around a hull, the airstreams divide into the right and left lungs.

The right lung is larger than the left because it isn't compressed by the bulk of the heart. The airway leading into it is more vertical too – if a peanut or a button is inhaled it's likely to fall into the right lung. From the lung root, where the great vessels enter and leave, to the leaf-like membranes at the periphery, the airways of the lung resemble a tree – specialists even use the term 'bronchial tree' to describe them. Their anatomy has been carefully worked out not just because children inhale little objects that have to be retrieved, but to aid surgery. If you want to chop out a lung tumour, root and branch, you have to remove the affected segment of lung as well as the branch of the airway that supplies it.

THE LYMPH NODE BIOPSY confirmed what I had feared: though the X-rays had originally been clear, Bill had lung cancer. The position of the tumour, and the fact that it had already spread, meant that surgery would not be an option. Doctors have an odd term for the quantity of cancer that builds up within an organ; they call it 'tumour burden'. As Bill's lungs became heavier, his body and voice became lighter, more insubstantial. At first he still managed to come down to my office to see me, but within a couple of months of his biopsy I was pedalling out to his house every fortnight or so to visit him. He remained as stoical as ever. During these meetings he'd usually have a cigarette in his hand, his nostrils billowing smoke like twin factory chimneys. He'd decided it was too late to bother quitting. As the smoke hovered in clouds over his head it seemed to give shape and substance to his words.

As the weeks went by his tumour grew, and as his lungs became heavier the sounds I heard in his chest began to change. I could hear the breath whistling at his carina, his voice crisply clear as it transmitted through his solidifying lung. It wasn't long before he needed supplementary oxygen to move around the house, delivered by little tubes worn over the ears and passing under the nose. Because oxygen is considered risky in the house of a smoker, he finally had a reason to quit the cigarettes. I asked him how difficult it had been to stop and he flashed one from his library of smiles. 'No bother at all,' he said, 'I should have done it years ago.'

He'd been diagnosed in the autumn, and by spring he'd had his easy chair taken out of the living room and a hospital bed installed. 'Fantastic, Doc,' he said with a grin as he showed me the electric buttons that would raise and lower it, sit him up straight or lie him flat. 'It's almost worth cancer to get hold of one of these.' He laughed, but his wife didn't.

There are landscapes, often of limestone, where tunnels within the earth truly breathe: they exhale in the heat of the day, and inhale as the earth cools by night. One afternoon, when I went in to check on him, his breath was like that: cold, slow and carrying the memory of having been deep underground. 'Look up there,' he said, pointing up the hill where a stand of trees was greening with the spring. 'Do you know what's through those trees?'

I followed the direction of his finger. 'No, from this angle I'm not sure.'

'The crematorium,' he said. After a few moments he added: 'I'm not scared. When I'm so breathless I can hardly move, I see smoke from its chimney and think it might not be so bad to be up there, blowing over the city.'

The wind was invisible, light as spirit, seen only in the motion of the trees and the smoke, which on that day at least was carrying the ashes of somebody else.

7

Heart: On Seagull Murmurs, Ebb & Flow

But hark! My pulse, like a soft drum
Beats my approach, tells thee I come

Bishop Henry King, '*Exequy*'

BEFORE STETHOSCOPES WERE INVENTED, physicians would listen to their patients' hearts by laying one ear directly onto the skin of the chest. We're accustomed to laying our heads against the breasts of our lovers, our parents or our children, but once or twice when I've rushed out on an urgent house call, leaving my stethoscope behind, I've had to rediscover the traditional method. It's an odd sensation – intimate yet detached – to apply your ear to the chest of a stranger. It helps if you stick a finger in the unoccupied ear. Once you tune out all the background noise you begin to hear the sound of blood as it makes its way through the chambers and valves of the heart. The classical belief was that blood travelled to the heart in order to be mixed with vital spirit, or *pneuma*, rarefied from the air by the lungs. The ancients must have imagined a churning within; air frothing with blood the way wind whips up waves on the sea. The first time I placed my ear to a patient's chest I was reminded of holding a conch shell as a child, listening to the imagined ocean within.

When any fluid is forced through a narrow opening there is turbulence, and just as a river flooding through a narrow canyon can be deafening, turbulence within the heart generates noise. Medical students are trained to listen very closely to the subtleties of those noises, and to infer from them how narrow – or obstructed – are the canyons of the heart. There are four valves in the human heart. When they close, you hear two separate sounds. The first sound is made as the two largest valves – the mitral and tricuspid – close at the same time during the active part of the beat (known as *systole*), when blood is forced out of the ventricles and into the arteries. These valves are so broad they have thick cords like harp strings attached to their cusps to reinforce them. The second sound is made by the other two valves – the pulmonary and aortic – as they prevent backflow while the ventricles refill (*diastole*). Healthy cardiac valves close with a soft percussive noise, like a gloved finger tapping on a leather-topped desk. If they are stiffened or incompetent there are additional sounds: murmurs that can be high-pitched or low, loud or soft, depending on the steepness of the pressure gradient across the diseased valve, and how turbulent the flow.

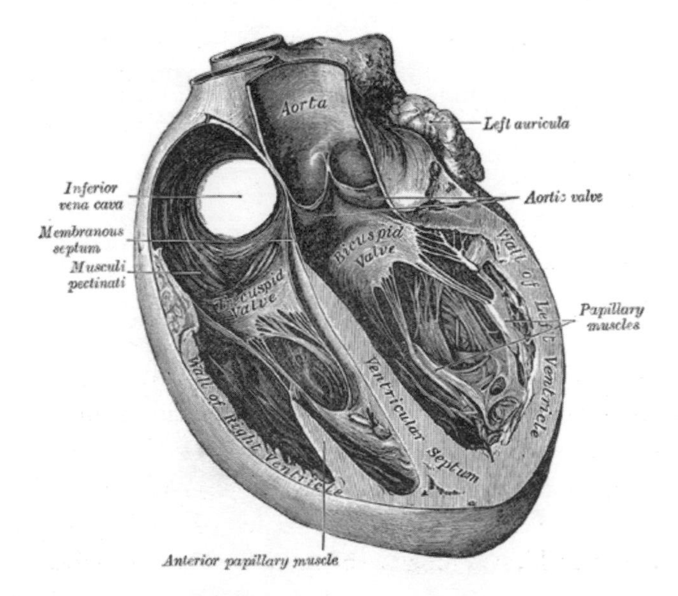

Starting out in medicine, I learned to tell the difference between valve pathologies by listening to a CD of murmurs. I'd put it on while studying, hoping that my subconscious would come to distinguish a 'seagull' from a 'musical' murmur, recognise the grate of mitral regurgitation from the trill of aortic stenosis. There was something comforting in listening to the gurgle of blood as I worked. I wondered if it recalled the sound of the sea, or hearing a storm outside while wrapped up warm, but the sounds were too rhythmic for that. Perhaps it's the womb, I thought, a deep memory of my mother's pulse.

It is the episodic squeezing of our heart, the pressure difference between systole and diastole, that gives rise to the pulses we feel in our wrists, our temples and our throats. The pulse is a familiar characteristic of life. Every so often someone comes up with a design for an artificial heart that pumps without need of a pulse. How would it feel, I wonder, to have blood that moved continuously through the body; not the ebb and flow of a tide, but a ceaseless, circular flow?

When language is called 'clinical' it is usually to imply that it is without emotion. Yet clinics are often awash in emotional transactions. In normal life it is unusual to see adults cry, but behind my closed office door it's routine. For the most part doctors are not emotionally cold, but become adept at shrugging off the burden of other people's misery. Clinical language has been drained of emotion not only because it's a shorthand between peers, but because it is a way of keeping patients' pain, disappointment and anguish at arm's length. Balancing empathy and compassion with a degree of detachment and professionalism takes experience as well as emotional intelligence, and no one gets it right every time. Hilary Mantel put it less generously, but more succinctly: 'Nurses and doctors are an elite, self-selected as sufficiently insensitive to get on with the job.'

The clinical language used to describe the loss of pulse

when the heart fails is not subtle. There may be 'rapid haemodynamic deterioration': the blood stops moving around the body. Presentation is with 'dyspnoea, syncope or pain in the praecordium': the patient gasps for breath and collapses, feeling as if their chest is being torn apart. People who suffer a complete valve failure, if they're conscious at all, have the conviction that they are about to die – and usually they're right. Doctors have a name even for this conviction and, like so much medical jargon, it's in Latin: *angor animi*, or 'anguish of the soul'. In the emergency room that feeling is taken seriously. I remember one woman I attended in the resuscitation room after she collapsed at her seventieth birthday party. As the nurses cut off her dress and pearls she gripped both of my forearms and pulled my face down to hers: 'Help me Doctor!' her eyes wide with horror, 'I'm dying.' Her pulse was impossible to find, and she died within minutes despite all our efforts to save her.

Since Descartes we've had a tendency to believe that from the chin down we are just meat and plumbing. *Angor animi* suggests that there is more to us than that; that in some way we become aware when a valve is no longer working or that a tear, or 'dissection', is developing within the wall of the aorta. As a sensation, *angor animi* carries great predictive power: I have ordered an urgent CT scan of the chest because of a patient's conviction that they're about to die.

It isn't just valve failure that can lead to sudden loss of a pulse: a blockage, or thrombosis, in the flow of blood through the coronary arteries can have the same effect. If the net of fibres coordinating ventricular contraction is starved of oxygen the heart muscle may begin to twitch chaotically or 'fibrillate'; death will follow quickly unless the muscle contractions are electrically shocked back into alignment. Some people remain prone to this fibrillation even after the blockage has been dissolved or forced open by a stent. Pacemakers have been developed that double as defibrillators – it is now possible to carry your own life-support machine, about the size and thickness of a Zippo lighter, tunnelled beneath a pocket of skin on the front of your chest. It nestles comfortably just under the collarbone. A patient of mine who had one, a war veteran, told me he wore it like a medal of honour. 'Mind you,' he said, 'when it fires it's as if a horse has kicked you back from the grave.'

Robin Robertson, the poet and editor, was born with a heart in which one of the valves – the aortic valve – was composed of only two cusps instead of the usual three. The aortic valve prevents backflow from the aorta into the principal ventricle of the heart. Each valve cusp consists of two elements: a firm nodule and a softer, more flexible crescent-shaped flap of tissue known as the lunula – 'little moon'. When a healthy valve closes, the three nodules snap together and support the flaps of the little moons, to control the tide of blood.

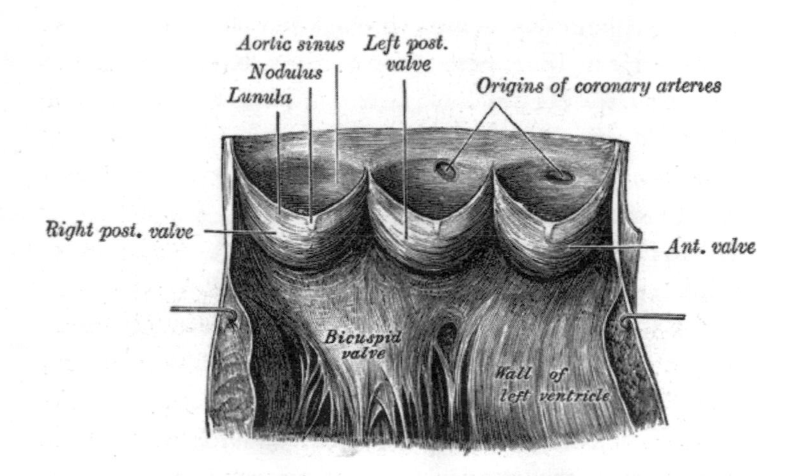

If there are only two cusps, not three, the lunules fit less snugly, which means blood begins to jet backwards into the ventricle. Sometimes this flurry of blood gets loud enough to be felt as well as seen; putting your hand flat against the sternum you can feel a fluttering, or 'thrill', through the leaking valve. For the first thirty years of Robertson's life those two cusps closed between seventy and one hundred times a minute, a hundred thousand or so times a day, around forty million times a year. Then he developed a 'seagull murmur', a name to describe the harshness of the sound, evocative of the squalls of turbulence that had begun to eddy in his heart. His poem 'The Halving' describes the operation he had to replace the valve.

In his poem Robertson describes how his heart was stopped and the circulation and oxygenation of his blood taken over by a machine. A carbon-coated disc, 'housed / expensively in a cage of tantalum', was taken from a sterile wrapping and stitched into his cross-clamped aorta. On waking from the surgery he felt disorientated and disembodied: 'Four hours I'd been away: out of my body. / Made to die then jerked back to the world.' Once the anaesthetics and morphine had drained from his bloodstream he was left with a pain that ripped through his sternum whenever he moved: raw bone grating on bone. When that began to ease, a paralysing darkness began to settle over his mood: 'Over the pain, a blackness rose and swelled; / "pump-head" is what some call it / – debris from the bypass machine / migrating to the brain.'

No one knows why some individuals experience 'pump-head': a disturbance of mood and cognition brought on by having your blood moved beyond the confines of the body, but a charge nurse in a cardiothoracic intensive care unit told me that up to a third of her patients experience it. Many are violent as they come round; security guards have to hold them down as they are sedated with powerful antipsychotic medication. Some are merely quiet, 'not themselves' as she put it to me: as if they have to grow re-accustomed to their bodies. Some become inappropriate and disinhibited; she told stories of vicars making ribald jokes and genteel ladies issuing foul-mouthed curses.

Some researchers think that when the aorta is cut, amputating the heart from its vessels, tiny fatty particles fly off into the arteries of the brain like flocks of birds, and are trapped there in a fine net of capillaries. Some believe that bubbles from the machine disturb the delicate balance of cerebral blood flow. Others have suggested that inflammatory processes within the brain, little understood, are set in motion by the trauma of having your chest prised open and your ribs wedged apart ('held aghast' as Robertson superbly puts it). Bypass machines cool the blood, and some think that 'pump-head' is a by-product of cooling the brain. But there is another theory: bypass machines have been in use for more than sixty years, but they still can't closely mimic a natural pulse from the heart. It may be that the heart's internal rhythm is essential to our well-being: our brains, and our sense of self, may depend on it.

It's nearly four hundred years since William Harvey realised that classical beliefs about the heart were wrong and that it works as a four-chambered pump. Before his *De Motu Cordis* was published in 1628, ideas hadn't moved on since Roman times. In fact, we still often speak as if classical beliefs were true, and the heart generated not only our pulse, but also our spirit. A heartless person is someone without conscience, even without a soul. We speak of heartache, and of following our heart's desire, of dying of a broken heart; we feel a conflict between our hearts and our minds, as if reason lay in the brain, but the heart were the helmsman. 'Pump-head' could be a manifestation of bubbles, of cooling, of fat, of inflammation of the brain, but to Robertson, the experience of having his heart stopped and his blood circulated through a machine was 'more interesting than that'. It left him feeling 'halved and unhelmed'; 'I have been away, I said to the ceiling, / and now I am not myself.'

Cardiopulmonary bypass machines have much in common with classical ideas about the function of the human heart: blood is drawn from the large veins in the chest then sucked into a chamber where it is able to absorb oxygen (or 'vital spirit'). The early machines bubbled oxygen through a reservoir of blood in the sort of churning Aristotle imagined took place in the ventricles. But since the mid-1970s we have suspected that it's better to keep blood and air apart, separated by a synthetic, disposable membrane.

Once it has passed through the oxygenator, the blood is squeezed through a tube by a roller, or drawn by a centrifugal pump. From there it is forced through a series of bubble filters and coolers, and then through sensors that analyse the blood for acidity, oxygenation and saltiness. It can be piped back into the body through a cut in the aorta just above the heart, but also at the carotid in the neck or the femoral artery in the groin. From the perspective of the human body as plumbing, it doesn't make any difference where you put it back in.

In the 1990s some prestigious scientific journals began publishing articles claiming that patients suffered less from 'pump-head' if the blood from the machine was delivered to them in heart-like pulses, rather than in a steady flow. Capillaries and cells perform the silent industry of life at the microscopic level; in the brain their function is intimately related to thought and personality. The evidence suggests that they prefer the blood that nourishes them to come in pulses. Even the best bypass machines manage only a clumsy approximation of the pressure pulse of a beating heart.

A FEW DAYS AFTER I heard Robertson read his poem a pregnant woman came to my clinic. She hadn't felt her baby move for a day or so, and wanted me to reassure her by listening for its heartbeat. Normal stethoscopes are no use for listening to the heartbeat of a baby in the womb; the sound is too fast, quiet and high-pitched. Midwives often use an electronic Doppler probe to find the foetal heart, but I used a modified tube called a Pinard stethoscope, like an old-fashioned ear trumpet, wedged between one ear and the swollen contour of the woman's belly. The best place to lay the trumpet end is where you think you've felt the convex curve of the baby's spine. Even with one finger in my other ear it took a while to find the heart – an agonising couple of minutes for the mother. But there it was: a rhapsodic, syncopated interleaving of her heartbeat with her baby's. The foetal heartbeat was distinct, fluttering fast like a bird over the oceanic swell of the mother's pulse. I paused for a moment listening to the two rhythms within one, two lives within one body.

*

'The Halving'
by Robin Robertson

General anaesthesia; a median sternotomy
achieved by sternal saw; the ribs
held aghast by retractor; the tubes
and cannulae drawing the blood
to the reservoir, and its bubbler;
the struggling aorta
cross-clamped, the heart
chilled and stopped and left to dry.
The incompetent bicuspid valve excised,
the new one – a carbon-coated disc, housed
expensively in a cage of tantalum –
is broken from its sterile pouch
then heavily implanted into the native heart,
bolstered, seated with sutures.
The aorta freed, the heart re-started.
The blood allowed back
after its time abroad

circulating in the machine.
The rib-spreader relaxed
and the plumbing removed, the breast-bone
lashed with sternal wires, the incision closed.

Four hours I'd been away: out of my body.
Made to die then jerked back to the world.
The distractions of delirium
came and went and then,
as the morphine drained, I was left with a split
chest that ground and grated on itself.
Over the pain, a blackness rose and swelled;
'pump-head' is what some call it
– debris from the bypass machine
migrating to the brain – but it felt
more interesting than that.
Halved and unhelmed,
I have been away, I said to the ceiling,
and now I am not myself.

8

Breast: Two Views on Healing

To be healed is not to be saved from mortality but rather, to be released back into it: we are returned to the wild, into possibilities for ageing and change.

Kathleen Jamie, *Frissure*

AN AWFUL DISEASE, breast cancer, afflicting young women and old, and so common that at any one time, most family doctors will know several women suffering it. In cutting a tumour out the breast is often disfigured and, in rich countries, cosmetic surgery is offered to mitigate any distress that the scarring – a perceived mutilation – might carry. Like the face, the breast is tangled up with ideas of beauty and youth: a reflection of our own anxieties about sex, ageing and loss of fertility. The cosmetic standards expected of breast surgeons are higher than in other specialties – fashion designers put breasts on display in ways that for other parts of the body would be unthinkable.

The special status of the breast carries over into the clinical management of its cancers: in the city where I work women with concerns about lumps in their breasts are generally seen more quickly than for other cancers, within a few days, and by the end of the clinic they will have seen a specialist, had the lump imaged by X-rays or ultrasound scans, and have had a piece of it removed for examination under the microscope if warranted. If someone has cancer then the options for surgery, chemotherapy and radiotherapy will often have been explained before she goes home.

Breast specialists are considered among the most approachable of surgical specialists; sensitive to the anxieties of their patients, and careful in their management and follow-up. But however emotionally aware clinicians might be, the places where they work are still clinics. There's a reason the word 'clinical' is synonymous with cool, detached efficiency. When I walk into my local hospital I don't feel it as a place of comfort and healing: its glass and steel facade, its labyrinth of whitewashed corridors, its sleek retail forecourt, are all reminiscent of a shopping mall, airport or exhibition centre. It's a place dedicated to the efficient processing of thousands of people; the hopes and anxieties of individuals tend to get drowned out in the crowd.

I LEARNED ABOUT DISEASES of the breast at the Western General Hospital in Edinburgh, which started out as a church poorhouse for the destitute in the 1860s. Some of the original poorhouse survives, nestled within a shell of modern buildings. As a junior doctor walking the corridors I'd notice the transition when the walls suddenly narrowed in, or the floor unexpectedly ramped a half-level. The hospital underwent further expansions through the Victorian and Edwardian periods, as the old poorhouse was transformed into a state-run hospital. The breast clinic was part of a later construction, built in the 1960s, when it seemed for a while as if the judicious application of science would deliver healing on an industrial scale.

Because the breast clinic was well funded it had some carpets and tasteful framed prints; the walls had been painted in pastel tones. But it was unmistakably a clinic: the waiting rooms had durable, wipeable chairs, and many of the rooms had no windows. I remember a surgical colleague leading me through an interconnecting series of rooms, introducing me to a series of anxious women who'd been referred because of lumps in their breasts.

Between one in ten and one in twenty of the women who attended would turn out to have a cancer; the others' lumps were all benign. Many had *fibroadenomas* – milk-producing tissue from the lobules of the breast that had become tangled in a complex web of ligaments and ducts. They are harmless apart from the worry they bring. Most of the others had *fibrocystic change* – a condition so common as to be considered normal. It's characterised by non-cancerous, fluid-filled cysts within the breast, which often wax and wane over the course of a menstrual period, like the phases of the moon.

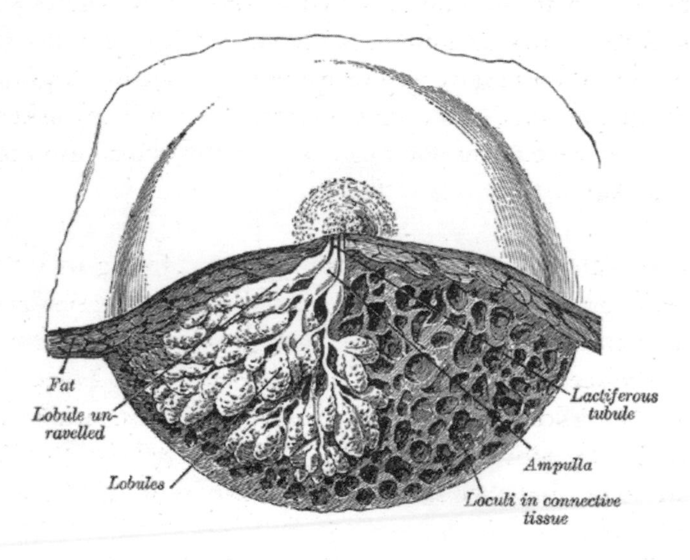

Those women with fibroadenomas would usually be reassured without arranging further tests – the lumps are characteristically painless, smooth and mobile, and much commoner in the young. Fibrocystic change can be more painful and difficult to diagnose, and so my colleague would perform a 'fine needle aspiration', passing a needle into each lump with the help of an ultrasound scanner, and drawing out amber fluid from each cyst. Occasionally she'd find a harder lump that seemed fixed to the tissues around it – a worrying feature. She'd use a broader needle to take a biopsy or, if it was too deep within the breast, arrange a 'lumpectomy' under general anaesthetic.

In the clinic we'd also see women who'd returned to have their surgical wounds examined to check they were healing properly. There were those who'd had mastectomies for cancer, others who'd had reconstructive surgery, and a few who'd had breast reductions because the weight of their breasts had begun to cause back strain. The patients were seen in rapid succession, having been allocated a cubicle and had their clothing arranged by one of the nurses to facilitate a quick examination. The wound was the focus of the consultation: how well or badly it was healing, and its cosmetic result. I don't remember the women being asked how they were coming to terms with the transformation of their bodies.

FROM A CLINICAL MANAGER'S perspective, healing may be seen as an impersonal, reproducible process, to be systematised and rolled out as cheaply as possible. One autumn I went to see a unique exhibition about breast cancer recovery that explored an alternative perspective: a collaboration between an artist and poet that examined one individual's path to healing once she'd got beyond those walls of glass and steel.

When, in her fiftieth year, the poet Kathleen Jamie discovered she had breast cancer, she didn't have immediate reconstructive surgery; after removal of the tumour she woke up with a long, Y-shaped scar that curled around her chest wall. It was a shock for her to look down and see her chest wall flattened in a way it hadn't been since childhood, her heartbeat fluttering beneath the skin. As she lay convalescing at home she began to think about her scar and the transformation it represented. A new line wound around her chest, and she observed that 'a line, in poetry, opens up possibilities within the language, and brings forth voice out of silence. What is the first thing an artist does, beginning a new work? He or she draws a line. And now I had a line, quite a line!'

She began with that line on the body, but began to see references in it to what's known, for better or worse, as the 'natural world'. It was a map, a river, a rose stem. She had been subject to the gaze of a great many clinicians through her treatment, and began to wonder how it would be to have the scar examined instead by an artist. She asked an artist friend, Brigid Collins, if she would consider making a series of paintings and sculptures relating to the scar. To introduce a measure of reciprocity into the project Jamie began to write short prose poems. Jamie's poems and Collins's artworks evolved in tandem, rather than being created to illustrate or explain one another – each of the women worked together to create separate but connected responses. 'The healing that took place therefore became a shared experience,' Collins later wrote, 'of both recent and past wounds, that can be seen at once as being both personal and universal, using our experiences of the natural world as a starting point.'

The exhibition the women put together had its origins in two separate traditions of visualising the body. The first considered anatomy through the lens of the surgeon-artists of old like Charles Bell, and the traditional medical illustrators, who prepared images of disease and mutilation for the purpose of medical education. Though beautifully executed those images were often amputated from their context – the lives and stories of the women they depicted.

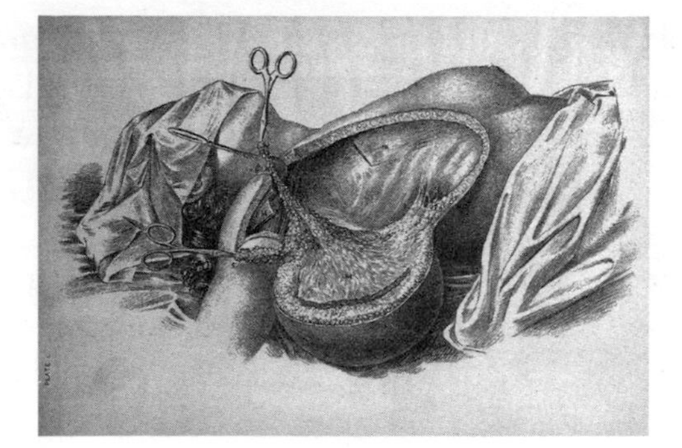

The second tradition they drew on was older, with its origins in classical perspectives on health, and imagined the body as a mirror of the cosmos. If the body is a landscape, and illness a disturbance in the greater harmony of which we are but a small part, then the world around us holds clues as to restoring inner balance.

When Jamie seems the tumour on her mammogram, it doesn't horrify her or symbolise a threat, but is instead 'rather beautiful, a grey-glowing circle, like the full moon seen through binoculars'. As she lies in her garden recuperating a flock of birds in the rowan trees recalls the image of that coagulated tissue: the birds are 'a density in their branches'. 'Sometimes I almost hear a sweet wild music,' she writes in one of her poems, 'it's audible in the space between the rowan leaves'. Sounds distant from the garden remind her of 'the sound of knots untying themselves, the sound of the benign indifference of the world'. An accompanying painting imagines her scar as a rowan branch, the text overlain by layers of gesso and shellac then sanded back to visibility, as if the text and the rowan leaves are emerging into new life. Another centres on a line from Robert Burns: 'You seize the flo'er, the bloom is shed', and inspired a painting of a dog rose, shaped to the contour of Jamie's scar and emerging from a stained page like the illumination of a medieval manuscript.

A feature of breast cancer is how often it runs in families, woman to woman through the generations. Jamie remembers sitting on her grandmother's knee as a girl, cuddling into her breast. 'My grandmother called her breast her "breist", her bosom her "kist"', she says in 'Heredity 2'. '"Come for a wee nurse aff her Nana," she'd say. "Courie in, hen."' The accompanying sculpture by Collins is called 'Kist'. It was designed as a 'place of safe-keeping', wrote Collins, 'a feminine embrace, a container, a sewing box, or skirt, which might keep one safe from the world'.

The last piece that Jamie and Collins created together imagines thousands of house- and sand-martins feeding over a river in preparation for autumn migration – they're 'kissing the river farewell' just as Jamie ends her summer of convalescence, 'preparing themselves, sensing in the shortening days a door they must dash through before it shuts'. A period of healing can be seen not just as something to be endured, but something to be thankful for: 'recovering from the operation was bliss of a sort', Jamie explained in her introduction to the work. 'No one wanted anything of me ... I walked by the river and slept better than I had in years.'

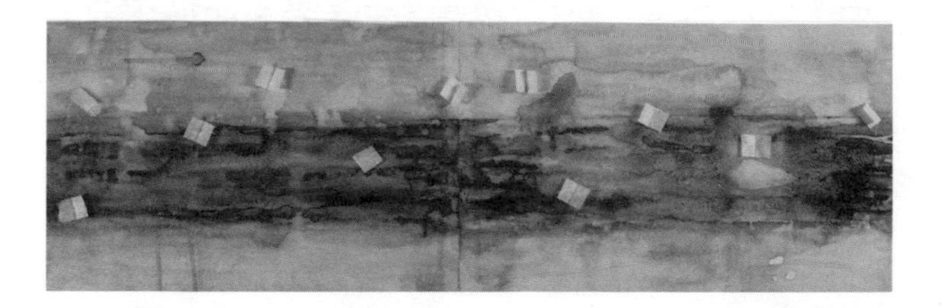

The name 'Frissure' was coined by Collins to describe the project. The scar is a fissure on the skin, and as Jamie explained, 'the naked, scarred body certainly causes a frisson'. The precise way Jamie has with language, and her lack of sentimentality, allowed her to take the anxiety and pain of breast cancer recovery and make of it a celebration.

When I was taken on a tour of the breast clinic, meeting half-dressed women in a succession of cubicles, it was because my teachers thought that that was the right way to learn about 'healing'. *Frissure* offered a lesson to take back into my medical practice – that healing involves restitution not just of our inner worlds, but an engagement with the environment that sustains us.

UPPER LIMB

9

Shoulder: Arms & Armour

But what are men, but leaves that drop
from their branches to the earth?

Apollo's speech, *The Iliad*, Book XXI, v 540

TRAINING IN EMERGENCY MEDICINE often felt like being awash in a sea of humanity; my pocket textbook a pilot-book for mariners. The departments themselves were often windowless as the engine room of a ship, and the staff moved in shifts just like deck officers on watch. Signing up for the training was a bit like enlisting in the Marines: the strict hierarchy of the medical staff, their bleached uniforms, their codes of behaviour, the alcoholic blowouts after hours.

On one afternoon shift it was sunny outside, but deep in the department there was only artificial light. A radio screamed an alert that an injured motorcyclist was on his way in by ambulance. The ambulance paramedic, Harry, let us know that although the biker was breathing and conscious his shoulder and chest had been badly injured. Harry was someone I'd come to know well in that department: battle-hardened, cynical but tremendously skilled at trauma life support.

A few minutes after the radio call, Harry hurried into

the room pushing the patient ahead of him. The biker had a moonish pallor to his face, and crew-cut black hair. I noticed first his rigid plastic collar, then his oxygen mask, then, with relief, that he was breathing for himself. Harry had slashed open the left sleeve of his leather jacket to fit a blood pressure cuff and an IV drip. He'd splinted the right arm because its position looked wrong – the right hand hung limply at an angle, like a snapped lance.

'Chris McTullom,' Harry said, 'twenty-five years old. He lost it on a bend, going forty or fifty I'd say. Hit the siding and went over the handlebars. There was a pillar by the side of the road – I reckon he went onto it with his shoulder.'

'How long did he lie?' I asked.

'Just ten or fifteen minutes.'

'Any sign he's lost blood?'

He shook his head. 'None. He's had a litre of fluid IV, blood pressure is a hundred over sixty, pulse is a hundred and ten – no wounds. He's a lucky lad.'

'Has he said anything yet?'

'Not much. Coma Scale is 11, pupils fine.'

I looked down at Chris and began to check him over: neck immobilised, breathing well and plenty of oxygen getting into his lungs. His pulse was fast but with good volume and there was no blood leaking onto the sheets.* His fingertips on the left were pink and warm. I yelled in his ear, 'CHRIS!', and his eyes opened, but then closed again. 'How is the bike?' he moaned suddenly, 'My bike …' He wouldn't squeeze my fingers when I asked him to, but when I pushed a pen hard down on his nail bed to check his responsiveness he pulled his hand away, swore and tried to punch me with his good arm. From being pale and expressionless his face began to boil with violence.

* You can bleed to death internally without a drop spilling onto the floor: pelvic fractures, femoral fractures or bleeding into the chest or abdomen can all cause enough internal blood loss to threaten life.

'GCS is 12 or 13 now – he seems to be coming round.'

McTullom was straining with anger now, trying to get up and off the table, but unable to for the pain in his arm and the restraints on his head and neck. With Harry's help I held him down and gave him an injection of morphine. He fell back into a doze, and we were able to cut through the protective armour of his jacket's right sleeve. There was no blood on his T-shirt, but his right shoulder looked distorted – instead of being muscled and square as it was on the left, it was a pulpy, swollen diagonal. Harry was right: he must have hit the pillar with his shoulder, slamming his weight onto the collarbone. Once he had been tranquillised by morphine we rolled him carefully on his left side while maintaining the straightness of his spine, to see if he had any other injuries of his vertebral column. All normal.

'Can you feel me touching your hand?' – I began stroking the fingers of his left hand. His teeth were gritted, but he tried to nod – an impossibility in a hard collar. 'Don't nod, just say uh-huh if you can feel me.'

'Uh-huh.'

'What about here?' I began to touch his fingers on the right. Nothing.

'And here?' I began to touch his arm higher up, towards the elbow, then the swollen shoulder. Nothing – he couldn't feel me touching the skin. 'Can you bend your fingers?' I asked, putting my own fingers into his right palm. There was a slight flicker as he tried to make a fist. 'Good. And bend your arm?' Nothing. The rage he'd shown just a few minutes before was starting to give way to a drowsy, drug-addled fear.

'What do you do for a living?' I asked him.

'Soldier,' he said. 'A gunner …'

When the X-rays came they showed that his right collarbone was smashed into pieces. There's a fine network of nerves behind the collarbone, emerging from the neck and controlling movement and giving sensation to the arm. He hadn't just broken up his shoulder in the crash; he'd paralysed his right arm.

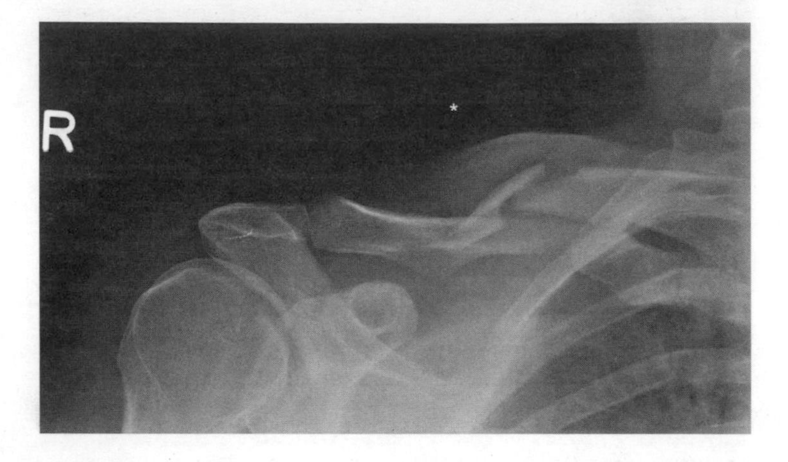

HUMAN CULTURE EVOLVES with the drama of history, but our anatomy, and the limitations it imposes on us, remains the same. Homer's *Iliad* was first written down almost three thousand years ago, describing a Greek siege of the city of Troy that may have taken place several centuries earlier than that. In Book VIII there's a scene of heavy fighting – Teucer the master-archer is bringing down a slew of Trojans, and being cheered on by his king, Agamemnon. 'I've shot eight arrows, and killed eight young warriors so far,' says Teucer, 'but there is one mad dog I cannot hit.' The 'mad dog' is Hector, a prince of the Trojans. The next passage is worth quoting in full:

> Hector jumped down from his chariot with a loud cry, picked up a great stone, and ran straight for Teucer in fury. Teucer took an arrow from his quiver and laid it upon his bow, but before he could take aim and fire Hector struck him with the mighty stone; he hit him on the collarbone, where it divides the neck from the chest – a deadly place. His hand and wrist were

> numbed by the blow, and as he fell forward onto his knees, the bow fell from his hand.

Teucer's brother Ajax ran forward and stood over the fallen man, shield aloft, to protect him from a rain of arrows. Two more of his comrades ran over and lifted him, 'groaning in pain' back to the safety of the Greek ships.

The author of *The Iliad* was a surprisingly accurate observer of anatomy. The battlefields of antiquity must have been chaotic places, sprawling with bodies and mired in blood. The warriors and camp-following poets were familiar with what is now called 'major trauma', and may have developed their own trauma care. There are some medically qualified Homer enthusiasts who have gone so far as to propose him as an early battlefield medic. Repeated through *The Iliad* are careful accounts of spear wounds, arrow strikes and sword blows, which take care not just to describe the part of the body that has been wounded, but the physiological effects of those wounds and, on occasion, specific treatments.*

When Hector paralyses Teucer's arm by hitting him 'where the collarbone divides the neck from the chest' it's an accurate description of a trick still used by martial arts experts today – 'The Brachial Stun'. A blow to this area may not just temporarily paralyse the arm: if it causes pressure on part of the carotid artery it can trigger a reflex slowing of the heart. In sensitive individuals the heart can slow to such a degree that the victim falls unconscious. There are innumerable 'brachial stuns' available to view on the Internet – home videos of US Marines practising on one another in their barracks, black belts filmed in the ring, even police officers

---

* But as the classicist K. B. Saunders noted dryly, 'I do not expect every wound described by Homer to be realistically explicable. One should try to come to some physical explanation of events if possible. But miraculous things do happen in *The Iliad* ... miraculous wounds should not be a surprise to us'; *Classical Quarterly* 49(2) (1999), 345–63.

attacking their suspects. Watching them, I thought of Teucer crumpling to the ground with his numb, lifeless arm.

THE NAME GIVEN to the spaghetti junction of nerves behind the collarbone is the 'brachial plexus', and when anatomy took a greater part in medical training every student had to memorise its arrangement:

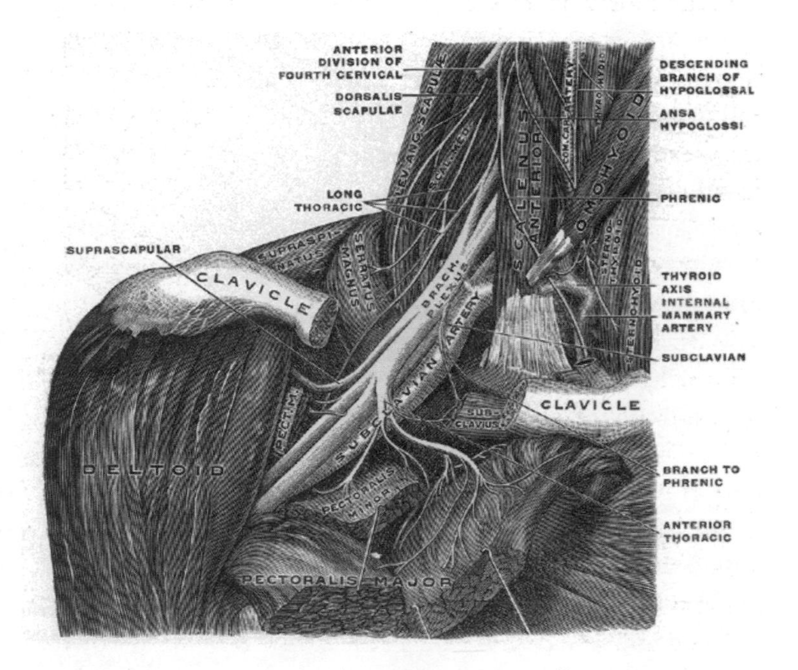

Five nerve roots from five vertebrae in the neck unite to form three 'trunks', which divide into anterior and posterior divisions. Those divisions perform an elegant interleaving with one another before braiding into three 'cords': 'medial', 'lateral' and 'posterior'. The posterior cord supplies those muscles which *straighten* the arm and wrist, as well as supply sensation to the back of the hand and forearm, while the medial and lateral cords activate those muscles which *flex* the biceps and wrist, and operate the small muscles of the hand.

The arrangement seems overly complicated, but arises from the way the arm forms in the womb. *Brachium* in Latin has the same root as our word *branch* – it starts as a bud, sprouting straight out from the trunk the way a branch grows from a tree. It begins to bud at just four weeks' gestation, and over three subsequent weeks divides into a rudimentary hand, forearm and upper arm, then rotates through ninety degrees. It's the movement of those muscles as the arm grows and rotates, and the fixed origin of the nerves in the neck, which provide the warp and weave of the brachial plexus. Homer didn't know of the origin of the plexus but he was acutely aware of its anatomy, and the martial advantage that knowledge could give.

BETWEEN TRAINING in emergency medicine and as a general practitioner I took a job as a medical officer in Antarctica. The British Antarctic Survey sent me as a ship's doctor to sail the length of the Atlantic Ocean and finish up at one of the most remote research stations in the world: Halley Base. The station would be isolated for ten months of the year in which I was to stay as base doctor. The evacuation of medical casualties was almost impossible for those ten months, so before taking up the post I was sent to a mixed military and civilian hospital for extra training.

The military doctors taught me how to give my own anaesthetics, drill out rotten teeth, and perform simple, single-handed trauma surgery. I'd always been suspicious of military medicine: to join a troop of soldiers intent on killing and maiming their enemies seemed to contradict every principle of ethical practice. Hippocrates said 'first, do no harm', but a close reading of his works also turns up 'he who would become a surgeon must first go to war'. From antiquity until today, war has provided an abundance of casualties to learn from: in medicine, as in other fields of expertise, practise makes perfect.

The military doctors taught me how to take my own

X-rays with a portable unit designed for the battlefield, reset broken bones and drill holes in the skull in case of coma following head injury – all skills they applied in war, but which they thought I might need in Antarctica. I was sent to traditional military establishments: dental anaesthesia at an air-force base, logistics at an infantry barracks. I took a course called 'Disaster Relief Operations' and sat in a room with thirty doctors, paramedics and nurses, all recently returned from battle zones. We learned how to build dressing stations near a front line, dig cholera-busting latrines, and other things that might be more useful on polar expeditions: satellite communications, improvised life support, and how to protect fragile drugs and equipment in transit. I developed an unanticipated respect for the military medics, and realised how much their predecessors had advanced our understanding of the body. Antiseptic surgery revolutionised the survival rates of soldiers during the Boer and First World wars, while the advent of antibiotics had a similar effect in the Second World War. Charles Bell had learned much attending the soldiers of Waterloo; the Roman surgeon Galen had been physician to the gladiators. Perhaps the anatomical knowledge shown in *The Iliad* was part of this long, often unacknowledged tradition.

THE WORD 'ARMS' is dual-use: parts of our bodies and weapons of war. 'Armed', 'armour', 'army' – our vocabulary bears witness to bodily violence, and humanity's attitude to killing is written into our figures of speech. Someone skilled in violence is often known as 'a strong arm', and soldiers with a common cause 'brothers in arms'. In Latin, *armus* means simply 'shoulder', while *arma* can mean any weapon, from a root meaning 'that which is fitted together'.

A historian of military medicine, P. B. Adamson, once read *The Iliad* with more care and attention than most surgeons bring to the closing of wounds. While acknowledging that it is an epic poem and not a historical record he noted every cut, together with the weapon that had dealt it and whether the wound turned out to be a fatal one. He then compared the results with a similar exercise on Virgil's *Aeneid* and concluded that at the time of the Trojan War spears were the most fatal weapons, but by the Roman period, which Virgil was describing, swords had the advantage. Stones were the least successful weapons in terms of killing people – 41 per cent of those hit by a stone end up dead (Teucer's life wasn't in danger when his arm was paralysed – after Hector disables him in Book VIII he pops up to fight again in Book XII). To be an archer like Paris or even Teucer, is, according to the subtext of *The Iliad*, to be slightly cowardly – archery delivers death from a distance as well as poorer accuracy: 74 per cent mortality as opposed to 100 per cent for swords and 97 per cent for spear thrusts. Adamson makes the point that in antiquity as today, armour encourages fierce engagement because it is reinforced towards the front but pitifully weak on the back. To turn on the battlefield and run has always been a mortally dangerous choice.

Adamson noticed that the legs are rarely injured in *The Iliad*, perhaps because the men often fought thigh-deep in the bodies of their fallen comrades, from the back of a waist-high chariot or even from within the protection of the hulls of their ships. He also notes that the head, neck and trunk are the parts of the body aimed for. When upper limbs are damaged in *The Iliad* it's usually because those arms are being raised in defence, or injured while they are themselves raised in violence. These Homeric patterns of injuries are still encountered every day in emergency departments: doctors assessing victims of domestic abuse often check women's forearms, as it is these that bear the brunt of warding off an attacker. A mid-shaft fracture of the ulna, the long bone of the forearm, is still known as a 'nightstick fracture' because it's most commonly encountered in those who've been beaten with a policeman's nightstick.

The pattern of wounds described by Homer remained broadly similar for almost three millennia after the siege of Troy – it was only after the widespread adoption of gunpowder, and the increasing distance between belligerents that it facilitated, that the pattern began to change. As weapons became more powerful, mortality figures paradoxically began to fall. Adamson compares the mortality and injury rates described in ancient texts with those that have been gathered from some of the most awful wars of the nineteenth and twentieth centuries.

Despite the appalling squalor and brutality of the Crimean War, the mortality rate from injuries was just 26 per cent – five and a half thousand deaths among twenty-one thousand British combatants. Proportions were similar for British troops in the First World War: of two and a quarter million soldiers, just under six hundred thousand died as a result of their injuries. Adamson shows that at their worst, shells and bombs turn up a mortality rate of 29 per cent (First World War), which is less than the rate for thrown stones described in *The Iliad*. The proportion of injuries sustained to the limbs versus those of trunk and head had entirely reversed: only 20 per cent of injuries in the ancient epics were to the limbs, but in the last century, injuries to the limbs make up 70 to 80 per cent of all those sustained in combat. As weapons grow more sophisticated, and kill at ever-greater distances, limbs started to become mutilated more often than soldiers were killed.

THERE ARE VARIOUS DEGREES of nerve injury. If the nerves behind the collarbone have been wrenched out of the spinal cord itself, there is almost no chance of recovery. If they've been ruptured there's a small chance that some may heal, and nerve transplants sometimes help regain some weaker function. Nerves are in some respects similar to copper wiring surrounded by plastic insulation sheathing: a nerve that has been severely stretched may regrow if its outer sheath has remained intact and only the inner 'axon', corresponding to the copper of the wire, has split.

Two months after his motorbike crash I saw Chris McTullom waiting in line for the neurosurgical review clinic. He still carried his right arm in a sling. The muscles of his upper arm that had been so pulpy and swollen were now withered and limp, but he had regained some movement in them.

'How are you getting on?' I asked him.

He took his arm from the sling, and slowly flexed his biceps. 'It's coming back,' he said. 'I'm not fit for duty yet, but perhaps in another couple of months.'

'And what then?' I asked him.

'Back to my unit,' he said. 'Afghanistan, probably.' He slowly curled the fingers of his right hand, stiff with disuse, as if to take hold of a trigger.

THE WORD 'ARM' might be embedded in our terms for weaponry and violence, but is also at the root of the language we use for friendship and affection. 'Embrace' means 'in arms'.

When the Greek and Trojan armies meet in Book VI of *The Iliad*, the Greek warrior Diomedes finds himself facing up to a Trojan named Glaucus, dressed in such magnificent armour that Diomedes thinks he must be one of the gods. 'What great man are you, among us mortals?' he shouts across the battlefield. 'At the threat of my long-shadowed spear you show yourself braver than the rest.'

'Why ask about my parentage?' Glaucus shouts back. 'Men are like leaves, they fall to the ground when their season ends, and spring brings new buds on the trees. So the generations of men die but new generations come to take their place.'

But after having initially refused to name his parents, Glaucus goes on to describe his ancestry: he is of Greek lineage; his grandfather was driven from Greece many years ago and settled in the lands of the Trojans. Diomedes realises that his own grandfather and Glaucus' grandfather had been friends, and because of that friendship he resolves to make peace: 'Let's stay away from one another's spears in the battle – there are many more Trojans for me to slaughter if the gods let me outrun them, and many Greeks for you to slay if you can.'

Standing apart from the hell of death that surrounded them, the two men leapt down from their chariots and clasped arms.

10

Wrist & Hand: Punched, Cut & Crucified

and (glancing at my own thin, veinèd wrist)
In such a little tremor of the blood
The whole strong clamour of a vehement soul
Doth utter itself distinct

Elizabeth Barrett Browning, *Aurora Leigh*

SATURDAY NIGHT SHIFT in the emergency department: payday weekend. The double doors onto the street have been like a storm drain, all the madness and misery of humanity pouring through them. At the end of my shift I navigate my way towards the changing room, between the old ladies on gurneys and queuing paramedics, handcuffed prisoners and policemen. Ambulance sirens are getting closer, a roar of shouts is coming from the waiting room, and from the noises in the resuscitation room I hear they're working on a cardiac arrest.

The changing room is windowless. Laundered green scrubs are stacked in piles on shelves, and a large bin of dirtied ones leans against one wall. The scrubs are made of some synthetic blood-proof cloth, and as they slide over my head they crackle with static electricity. I open my locker, throw in my name badge, and sift out my clothes from the discarded blood tubes, pens, surgical gloves and disposable scissors that have accumulated over the months. A colleague is changing into clean scrubs, beginning his ten-hour day shift. 'Good luck,' I say to him. 'You'll need it.'

Standing in the shower at home, scrubbing off the dried blood from my cheek and the smell of hospital disinfectant from my hands, I do a mental tally of the people I attended through the night: the overdosed and toxic; psychotic and broken; burned and convulsed. Seen from the corridors of an emergency department the world is mad, bad and, like the poet said, incorrigibly plural. 'How can you face it?' a friend asked me. 'So many of the people you see must have brought their misery on themselves.' Does that matter? I remember thinking. Few of us manage to be who we aspire to be. I like that in the emergency department life is extreme and unfiltered: there is no preferential treatment for those with power and money. Everyone sits together on the same hard plastic chairs, and is stitched up and in the same curtained cubicles. There is an unarguable democracy to 'triage': being prioritised on the basis of medical need, rather than influence.

Once out of the shower I notice it is 9 a.m. and tumble into bed the way a shipwrecked sailor would throw himself onto a beach. There are eight hours before I have to go back. The shifts come in a relentless tide: fourteen-hour night shifts, ten-hour day shifts, a couple of days off then straight back to the nights. All the time I work in adult emergency medicine I reverse my body clock through twenty-four hours every week or so.

The idea behind my training there was to learn how to approach every injury and intoxication that humanity can inflict on itself, but what I didn't bargain for were the stories. As I collapse into bed, my body twitching with fatigue, my neck and shoulders already tense at the thought of the next shift, it is those stories that keep me from sleep.

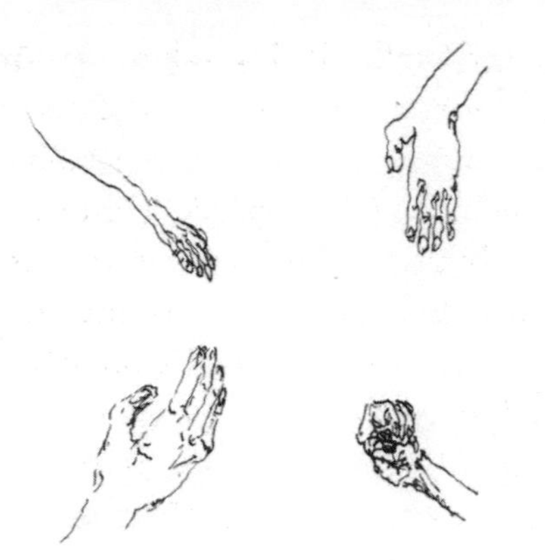

A MAN LIES on a gurney trembling, a hospital gown over his legs and chest. Beneath the pressed institutional cotton his body has a toned, athletic form; well tanned with the musculature of someone who doesn't waste his gym subscription. At the entrance to his cubicle I glance down at the clipboard: 'Mr Adrianson?' I say to him. He nods and I walk in, pulling the curtain closed behind me.

Tea towels are wrapped around his left forearm. Once a dirty white they are now a deep and lustrous scarlet. The topmost one, a souvenir from Majorca, has come partly undone and lies loosely at his elbow. Blood is pouring over his skin like a wet sunset, pooling in the crevice formed by his buttock and the rubber gurney mattress. 'I'm bleeding,' he says pointlessly as I reach to rewrap the arm with the towel, and begin to press hard.

'You're going to be fine,' I say, though I've no idea what's under the towels yet. Maybe he won't be; maybe the arteries are severed and the tendons too. Into the undamaged crook of his right elbow I push a 16-gauge cannula – as thick and long as a hat pin – pulling out the steel introducer as I nudge in its clear plastic conduit. Once the plastic wings of the cannula are taped down I draw off blood samples for haemoglobin and crossmatching, then hook up an IV drip of plasma substitute. 'Are you left-handed?' I ask him. He nods. 'What's your job?'

'I'm a pickpocket,' he says with a wry smile, 'what's it to you?'

'Just checking you're not a concert pianist.'

'I fell through a window,' he says and looks away, though the nurses have already told me another story. When the paramedics arrived at his house there was a woman sobbing in the corner, who told them that he'd been about to punch her but punched a door instead. The window panels of the door shattered badly, and I wonder if he has fractured the bones of his hand in the punch. As I press on the forearm I lift his hand and glance at his fingertips: nice and pink, so there's plenty of blood still getting down to them. I press hard on the pulp of his thumb, release the pressure, and count the number of seconds it takes to pink up. It's less than two, so inwardly I relax a little. The knuckles are in bad shape though, and as expected his little finger looks shorter than it should, and turned in at an unnatural angle. He has snapped the bone in the hand that supports it: a 'boxer's fracture'.

As I push on the forearm, trying to get the oozing to stop, I'm thinking of another boxer's fracture I dealt with earlier in the week. The metacarpal in question had belonged to a prison warder's fist, and only moments before assessing him I'd diagnosed his prisoner with a broken jaw. The two men sat in adjacent cubicles. The connection between the injuries was so obvious that it seemed almost discourteous to mention it. The warder had told me he'd been interrogating the prisoner about a disturbance and had his hands over the back of a chair, when the prisoner had kicked a desk which slid across the floor and made a bullseye collision with his knuckles. 'Is there any other way you can get a fracture like this?' he had asked me, nervously.

'No,' I'd said firmly. 'It's called a boxer's fracture. It happens when you punch something harder than the bones of your fist – or someone.'

The blood wells up more slowly now, so I pull back the tea towel and peep underneath. There is a long gouge in his forearm extending onto the wrist, as if he'd been mauled by a lion. And within the wound lie his muscles and tendons, glistening.

The nurses had already ordered an X-ray, and from looking at it I know that there is a sickle-shaped spicule of glass embedded somewhere in the wound. I elevate the skin around that wound now, dabbing with gauze and looking for the piece of glass. At last I find it, by touch rather than by sight, marbled with strings of clotting blood and tearing into the tissues like a poisoned thorn. I hold up the shard to the strip light and then walk over to the light box where the X-ray images are displayed. The bones of the forearm – the radius and ulna – are outlined in ghostly elegance as if etched on glass. I can see that his fifth metacarpal, the bone that supports the little finger within the heel of the hand, is fractured but not so badly that I'll have to twist it straight. I hold the shard up to the sickle-shaped opacity on the light box and find that the two shapes match one another completely.

'Good news,' I tell Adrianson. 'There are no more bits of glass.'

I sit down at the side of his trolley, and look down on the muscles of his forearm as they gather towards the wrist. The tendons of the superficial finger flexors glint in the light: the thick bands of collagen are like the quills of a feather, but in place of the barbs and vanes of a feather are fleshy chevrons of muscle. I ask him to flex his fingers, and marvel at the sight of the muscles bunching – the extraordinary intricacy of the pulley systems that control the fingers. *How mechanical we are*. The tendons are all intact; he can grip my fingers as strongly on the left as on the right, and I can't see any nicks in the surface of the tendons as they move in and out of view.

'When can I get home?' he asks.

'Just as soon as I've stitched these wounds and strapped up your broken finger.'

As a doctor I talk all day, taking histories and giving explanations. Sometimes I get to the end of a shift or a clinic and feel the need to be silent for hours, just to restore a balance. The verbal process of diagnosis works through sieves of possibilities, question and answer, weighing and measuring the patient's responses and deciding when to question further, and when to move on. It's a skill that takes years to develop: a medical history can take a student an hour, but as a GP or hospital consultant we have to try to make a decision within minutes. Practical tasks like stitching wounds or putting plaster of Paris on a broken limb offer a rare opportunity to spend time talking with a patient without that urgency; without directing the conversation towards a goal. There's a deep pleasure in performing a skill that's purely technical, with little of the intellect involved. Stitching is a technique, and like all techniques it can be done well or it can be done badly. Doing it well requires a level of focus that comes as a relief after the constant distractions of the emergency room floor.

I set up a sterile tray of instruments and suture thread, syringes of local anaesthetic, swab out his wounds again with antiseptic, and begin to stitch. He might need thirty or forty sutures, so this could take a while.

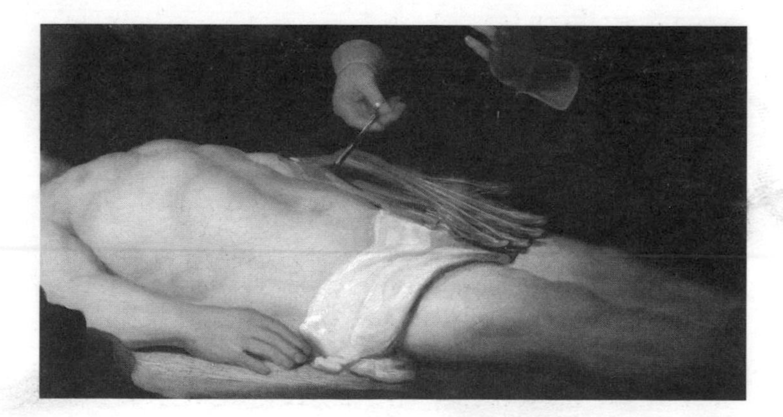

In the emergency room I've never seen someone die from slashing open the arteries of the wrist – generally they don't bleed enough to risk death. The only person I've seen die after slashing her radial artery had also taken a knife to her own throat, and managed to cut her carotids as well. Arteries are only two or three millimetres wide at the wrist, and when they're sliced open they often close themselves off as if in self-defence. But I have seen hundreds who scratch and cut their wrists not necessarily through a desire to die, but in an attempt to relieve extreme personal anguish, and demonstrate their rejection of the life they're obliged to live.

Slashing at your wrists is a way of lashing out at life: through the pulse, the wrist is emblematic of life, testifying to the strength and vitality within. It's a common way of giving release to feelings of tension: up to 4 per cent of the population admits to self-cutting (what is known as 'deliberate self-harm' or DSH), and though the wrist is the most popular, forearms, legs and hips are also common. Teenagers admit to much higher proportions, around 15 per cent, with girls more likely to come forward for help than boys. Cutting is often precipitated by feelings of extreme anxiety or distress, temporarily relieved by the act of drawing blood. As one self-harmer explained: 'As the blood flows down the sink, so does the anger and anguish.' An anthropologist who has studied self-harming behaviour called it 'a strategy of withdrawal or self-abasement used to show those one must both love and obey that one is hurt by them'.

The self-harmers I see are often teenage girls who are placed in impossible situations: pulled between the expectations of their parents, the demands of their peers, and an anguish that's partly about grieving their childhood, and partly about finding an adult identity. Cutting conveys the depth of conflict they feel, showing their families and friends just how appalling they feel inside. 'Communication of emotional pain to others may result in validation of that pain', wrote one group of DSH researchers, 'and demonstration of the severity of problems may elicit help, or maintain a valuable relationship.' From this perspective, to cut yourself is a rational decision.*

For the most part, the teenage girls I see haven't suffered systematic, tormenting abuse at the hands of those who were supposed to look after them, but childhood abuse is often the precursor of such cutting: having been abused as a child quadruples your chances of self-harming as an adult. When I meet people who self-harm in clinic, I try to elicit whether they have been, or are being, abused, but how likely they are to admit that to me, I don't know.

IN THE EMERGENCY DEPARTMENT there's a 'psych cubicle': a room with more privacy than the usual cloth-curtained spaces, and stripped of anything which could be turned into a weapon. It's telling that the room in which we assess patients who are mentally ill is the same room as that reserved for prisoners. It has two doors, so that a patient can't get between you and your exit, and both of them are lockable.

Melissa wore cheap plastic trainers, stained pink jogging bottoms, and a shapeless pink pullover with 'Gorgeous' written across it. Her hair was unwashed, bourbon-brown, and her eyes were liquid with panic. I'd picked up her file on the wall outside – it had her name, date of birth and the address of some nearby supported accommodation: a place where those with severe mental health problems can live quasi-independently, helped by trained staff and social

* One strategy to reduce the scarring of DSH is to encourage those who do it to hold ice cubes against the skin until it hurts instead, or sting their skin by stretching and releasing an elastic band placed around the wrist.

workers. Across the top of her file the triage nurse had written just 'DSH'.

She sat in the psych cubicle looking at the floor, checking and rechecking the dressings on her forearms. The sleeves of her top were pushed up to the elbows to make them more visible. She had five or six adhesive dressings on each forearm, and spreading out from their margins I could see old scars: the skin surface was ridged and fissured as unpolished marble.

'It's because I was abused,' was the first thing she said. I nodded.

'That's awful,' I said. At times it's the only thing to say.

'It was my grandpa – he's dead now – got what he deserved.'

She had been cutting herself only half an hour before and the blood was still spreading through the dressings.

'I didn't stop it. I should have stopped it. I'm so stupid.'

I sighed, and shook my head. 'How old were you when it started?'

She shrugged. 'Two? Three?'

'So you were just a tiny girl, how could you have stopped it? It wasn't your fault.' We sat in silence for a few moments together. Outside I could hear the clatter of trolleys and the arrival of ambulance sirens. 'What tablets are you on?'

'I don't want any tablets.'

'Are you sleeping?'

'Not for three days.'

'Well I could give you something to sleep at least, and let you rest.'

She nodded her head.

'Will you let me take a look at your cuts?'

She nodded, and held out both forearms. I began peeling off the dressings: the cuts were just shallow grazes, not deep enough even to need paper butterfly stitches, far less needle and suture. Slowly I began washing the cuts, and covering them with fresh dressings.

'You got down here to the hospital on your own, that was well done,' I said. 'You knew when you needed to get help.'

With teenage girls, sometimes just having their cutting acknowledged by those around them is enough – the habit stops when the family around change their own attitudes, or the girl grows old enough that the tensions of adolescence begin to resolve. Melissa's anguish had far more sinister origins; I felt utterly powerless to help.

ANOTHER WEEKEND NIGHT, so busy the patients are queuing outside the waiting room and along the corridor. There's a six-hour wait to be seen. At the nurses' station there's a radio tuned into the ambulance system; the police and the paramedics use it to alert the department when multiple, or very serious, casualties are on their way. It rings: a sound like a klaxon that makes even the most experienced staff jump.

'Major RTA on the city bypass,' says a voice in the radio, and requests an ambulance fitted out to carry two doctors to the accident scene. The ambulance staff don't request it often because it takes two doctors from the emergency room floor, but if there are casualties trapped in a vehicle, then calling it out can save lives.

I won't be going; I'm allocated to the minor injuries area for the night. But with only five doctors on the floor instead of seven the waiting time is going to get even longer. Braced for the fury that is about to break over me, I stand at the doorway of the waiting room to tell the patients.

'At the moment it's six hours to be seen,' I shout out, 'but two doctors have just been called away to deal with another emergency, so it will take longer. If you think you can go home for the night and be seen tomorrow, please come forward.'

The waiting room goes silent; everyone sits tight and glowers at me. In the front row I can see a girl with a bag of frozen peas on her ankle, a man holding a cloth over his eye, an old lady with a graze on her forehead – but each has been waiting some hours already, and no one wants to get up first. After a few moments, a man at the back wearing a boiler suit and work boots gets up. He's young – in his early thirties – with long sideburns and a splendid keel of a nose. His hand is wrapped in an old beach towel. 'I can probably come back tomorrow,' he says. As he speaks, his Adam's apple bobs up and down like a float.

I take him into the adjacent cubicle. His tells me his name is Francis, and as I unwrap the towel I jump back: there's a nail through his palm.

'There's a nail through your palm,' I say, pointlessly.

'I know.'

'What happened?'

'I was working late on the house, was getting tired ... and fired the nail gun by mistake.' The nail is clean, about four inches long; the puncture wounds on each side are neat with a halo of dried blood. He laughs: 'I was lucky it didn't fire right into wood,' he laughed, 'or I might still be there, pinned to a beam like Jesus.'

WITHIN THE PALM of the hand are four bones – the metacarpals – one for each finger. A fifth supports the base of the thumb. Between each bone are the delicate nerves that supply sensitivity to the fingers, some blood vessels, and also the muscles that splay the fingers or bring them tightly together (the muscles which bend or straighten the fingers lie in the forearm, not the hand). The metacarpal bases are bound to the bones of the wrist by tough ligaments, but further out, towards the fingers themselves, they are held fairly loosely. It's quite possible to fire a nail through the palm of your hand without causing any major damage: the nerves are narrow and run close to the bone, and the main blood vessels run in a broad arch from the heel of the hand to the base of the thumb, away from the palm itself. Firing a nail through the wrist is a different matter: the wrist has a tight, seed-like intricacy of nerves, blood vessels and interlocking bones.

Francis might have joked about crucifixion, but if you wanted to nail someone to a piece of wood, you wouldn't do it through the palm of the hand. The same anatomical features that allow a nail to pass through without causing serious damage mean that the structures of the hand aren't strong enough to support the body's weight. The tissues would rip and your hand would come free – mutilated and useless, but free.

Francis's fingers were all flexing normally and his sensation was undamaged: none of his nerves or tendons were hit by the nail. The blood flowed to his fingers as it should. The X-ray of his hand showed the nail passing beautifully between the metacarpal bones, as if shot through the bars of a cage.

After cleaning up his wounds I sent him to the plastic surgeons. They would pull out the nail in an operating theatre, in order to have a proper look into the hole and make sure that no fragments had been left behind. No matter how neatly they closed up the wound, he'd be left with stigmata on either side of his hand; a lifelong reminder of the night he was almost nailed to a beam.

IN THE 1930S a zealous French surgeon called Pierre Barbet became passionately fascinated by the details of crucifixion. To test whether the hand could support the body's weight he experimented by nailing cadavers to a wooden cross. Making a guess at Jesus's weight and the position of the arms with respect to the torso during Roman crucifixion, he calculated that the nails must have been hammered through the small bones of the wrist rather than the palm. Those wrist bones – the 'carpus' – are held together very tightly by ligaments; Barbet found that if he nailed his corpses by the wrists rather than the palms, they didn't tear out.

Pierre Barbet published his experiments on the nailing of a human body in the 1930s, but in 1968, in a burial cave near Jerusalem, a skeleton was found of a young man who'd been crucified during the Roman period. A nail about eleven centimetres long had been driven into the outside aspect of his right heel bone – the calcaneum – and traces of coarse olive wood, presumably used in the vertical stake of the crucifixion, were found under the head of the nail.

Dramatic claims were made after the find – the first direct evidence of Roman crucifixion – and the professor of anatomy at the Hebrew University suggested that a single nail had been put through both feet, that the forearms had been nailed, and that the victim's legs had been broken, while still alive, in a *coup de grâce*. Fifteen years later two sceptical colleagues – Joseph Zias and Eliezer Sekeles – re-examined the remains and came to different conclusions: the nail had been passed through only one heel – the right (the other heel bone had been lost) and the arms showed no trace that they'd been nailed. They concluded that crucifixion, as practised by the Romans, involved tying the arms to a T-shaped cross-beam with *rope*, and nailing each heel to a vertical stave. Olive trees usually generate straight beams for only two to three metres at the most, and so victims would not have been hoisted very high.

That Roman crucifixion occurred through the palms is such a commonplace in Western culture that 'stigmata', the development of bleeding wounds over the points of the body where Jesus was said to have been nailed, have surfaced throughout the last millennium. I've read of them on palms, wrists, the flank (where Jesus was said to have been stabbed), and even on the tops of the feet. I haven't heard of them happening on the side of the heel, and I'm yet to see someone fire a nail gun through their calcaneum.

ABDOMEN

11

Kidney: The Ultimate Gift

Nowadays it is possible to say that lives are connected, by transplant, across the thresholds of life and death.

Alec Finlay, *Taigh – A Wilding Garden*

IN THE INDIAN FOOTHILLS of the Himalayas there's a Tibetan hospital that serves the community around the home of the Dalai Lama. Between training in emergency medicine and beginning in general practice I worked there for a few months, managing the leprosy, dog bites, tuberculosis, dysentery and injuries of the local Tibetan population. It was a general hospital that turned no one away, and the job involved delivering a lot of babies and looking after two wards full of patients, as well as outpatient clinics twice a week. Through translators I'd labour to understand fifty or sixty newly arrived refugees, most of who were suffering from stress headaches, indigestion, homesickness or diarrhoea. Occasionally there'd be a forlorn westerner in the queue, pale and emaciated with dysentery they'd picked up by drinking unfiltered water. 'I want to live like the locals,' they'd say; I'd inform them the locals got dysentery too.

There was an alternative to the hospital: just down the road was the Tibetan Medical and Astrological Institute. Traditional Tibetan medicine is an ancient system involving the manipulation of five elements and three humours – practices resonant with Vedic and Hippocratic perspectives on the body. Those patients with vague aches, and unusual constellations of symptoms that we couldn't make sense of, often did well with the traditional Tibetan physicians. I often wish I had a comparable clinic down the hill from my office in Scotland.

Out of curiosity I visited the Institute, a grand white-washed building set among pine trees, on the spine of a ridge coming down from the Himalayas. Great charts of the human body were hung on the walls inside, overlain with meridians and lattices of lines, like the contours and grid squares on a map. Sometimes I understood the rationale for a particular Tibetan treatment, but for the most part it was a mystery – my understanding of the body didn't concord with theirs at all. If the kidneys weren't working, for example, the traditional practitioners thought that it was because the organs were too cold. The diagnosis 'cold kidney' was an illness all to itself, called 'k'eldrang'. Treatment of k'eldrang involved the avoidance of cold or wet seats, strains to the back, and certain foods thought dangerous for their cooling properties. In severe cases 'moxibustion' was recommended: an ancient practice with its roots in Chinese medicine, which uses burning herbs to heat the skin over particular meridians.

Tibetan customs of pilgrimage include carrying stones from place to place over the landscape. It's a practice I recognised from Scotland, where walkers often leave stones on the high ground of a particularly difficult or exhilarating climb. Once, when visiting a Tibetan monastery's prayer rooms, I saw an old monk touching a pilgrim on the head and back with a special stone – it was smooth, dark and shaped like a kidney. I asked what was being done. The stones can heal, I was told; being touched by them can rebalance the flow of energy within the body.

Traditional Tibetan medicine seemed to have some

success, but I was doubtful that sacred stones could be successful against kidney disease or renal failure.

THE WESTERN UNDERSTANDING of the kidney was slow in coming. Kidneys strain urine from blood, even Aristotle knew that, but as late as the fifteenth century one of the great Renaissance anatomists, Gabriele de Zerbis, still thought that the upper half of the kidney gathers blood, then strains it through a membrane strung across the middle of the organ. Anatomists like him had cut through human kidneys and can't have seen any such membrane, because it isn't there. Perhaps they *wanted* to believe in the existence of one so much that they saw it.

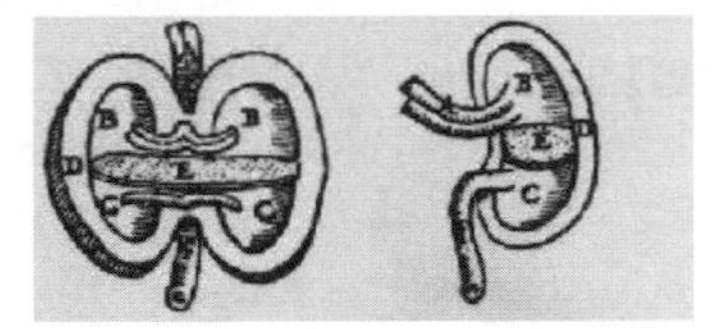

De Zerbis was a professor at Padua in north-east Italy, and wrote one of the first treatises on the medicine of old age – *Gerentocomia* – in the late fifteenth century. To retard the advance of old age he advised living somewhere with an easterly exposure (north-east Italy perhaps?), plenty of fresh air, and to eat a combination of viper meat, a distillate of human blood, and a concoction of ground-up gold with precious stones. Esteemed throughout the eastern Mediterranean as a specialist in medical care of the elderly, de Zerbis was called to Constantinople in 1505 to treat a member of the Ottoman elite. The old Ottoman died, and so de Zerbis was caught, tortured and sawn in half, just like one of his dissected kidneys.

De Zerbis's successor at Padua was Vesalius, a Dutchman who effected a revolution in anatomy and medicine (in those days, there was little distinction between the two).

Vesalius took the innovative step of describing what he *saw*, rather than what the textbooks, some of them dating back to Roman times, told him he *should* see. He cut kidneys in half and saw no membrane. He still thought that kidneys filter blood in some way; he just admitted he didn't know how they did it.

No one would come closer to the true mechanism until microscopes became commonplace a hundred and fifty years later, following advances in lens and prism technology. In the 1660s, lenses were achieving transformations in the understanding of both inner and outer space: near Cambridge, Isaac Newton, in quarantine from the plague, used his time to demonstrate how sunlight can be broken into colours by a prism, and formulated his laws of gravity. In London, Robert Hooke published his *Micrographia,* which showed the astonishing intricacy of tiny, everyday structures, such as body lice, pieces of cork, and flies' eyes (he coined the word 'cell' as the basic unit of life, because under the microscope they resembled a series of monks' cells). Around the same time, the professor of medicine at Pisa, Marcello Malpighi, used the microscope to demonstrate how blood and air did not mix freely in the lungs, but were merely brought closely together. He also revealed how capillaries in the kidney formed tiny sieve-like structures. He saw that the pale, central portion of the kidney was composed of masses of tubules; when squeezed these tubules produced a liquid that tasted just like urine (before biochemistry labs, the analysis of substances was often left to the tongue).

It took another two hundred and fifty years – until the early twentieth century – to understand the function of the kidney: the way renal blood vessels form a knot of capillaries which filter toxins into a cup-like receptacle at the head of each tubule. As vital functions go, it's one of the simplest that the body performs, but even so the subtleties of the process have proved fiendishly difficult to understand.

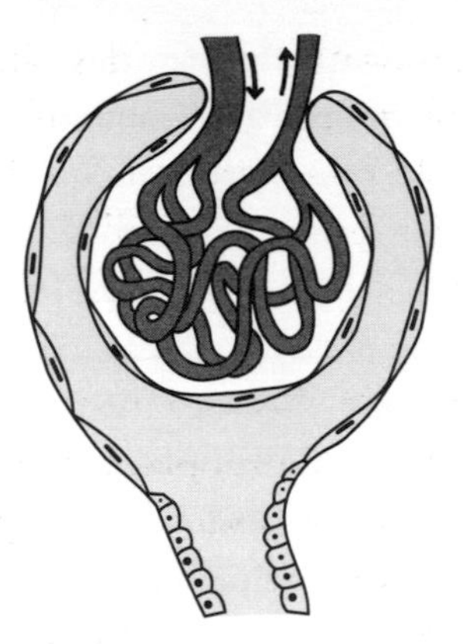

The function of the kidney seemed enticingly simple to replicate: the first attempt at building an artificial one took place as early as 1913. The machine was tried on dogs, with an extract of ground-up leeches used to prevent blood clotting inside it. Thirty years later, a Dutch physician, Willem Kolff, invented the first functioning kidney 'dialysis' machine for humans, which would artificially filter toxins from the blood – he didn't patent his machine, because he wanted others to develop it and make it more widely available.

Kolff initially worked under the scrutiny of Nazi occupation, but in secret he was a member of the Resistance. His first machine used the newly invented cellophane from sausage manufacturers, orange-juice tins and a water pump he obtained from a Ford dealer, but he refined it sufficiently so that in 1945 his machine saved the life of a sixty-seven-year-old woman. In 1950 he emigrated to the US and developed the process even further. As he was working on his dialysis machine, and more patients with kidney failure were beginning to benefit from it, something near miraculous happened: the successful transplant of a kidney from one body to another.

The apparent simplicity of kidney function led to the idea of building an artificial one, and the simplicity of kidney anatomy – one artery and one vein, and only one outflow for urine – meant that it was the first whole organ to be considered a candidate for transplant. The first kidney transplant in humans was attempted in 1951, but failed because the immune system of the recipient rejected the 'foreign tissue' of the donor's kidney. In 1954, at Brigham Hospital in Boston, this problem was circumvented by transplanting a kidney between identical twins, one of whom had suffered double kidney failure. The recipient's body was genetically identical to the donor, and so there was no rejection. It was the first time in history that an organ had successfully switched bodies.* The next twenty years saw a tremendous advance in the understanding of the immune system, and how to improve the recipient's tolerance of the foreign, transplanted tissue. By the late 1970s such operations between genetically dissimilar individuals were almost commonplace.

BRAIN TISSUE can only survive for a few seconds without blood, but kidney tissue is much more resilient – if kept cold an extracted kidney can survive for twelve hours or more (though the quicker it is transplanted the better). This means that kidneys can be taken from someone recently deceased or brain-dead, or even from a live donor, many hundreds of miles from where someone is waiting to receive it. National databanks now match recipients to available kidneys; the immunological profile of each is compared so that the chances of rejection are minimised. The kidney for the first transplant I saw had arrived by air from a city three hundred miles away. Its former owner had died that morning, and it was transported to theatre in a cooled polystyrene box.

Between the surgeon and myself lay Ricky Hennick, a man in his thirties who had suffered total kidney failure many

* The transplantation of skin had already demonstrated to surgeons that the transfer of tissue between identical twins was tolerated without 'rejection' by the recipient.

years before as a result of infections. He'd been kept alive during those years by dialysis. Only his lower abdomen was visible between the piles of green drapes; he'd been cut not in the back, where his own scarred kidneys lay, but on the lower left of his belly, making an opening into a cavity called the 'left iliac fossa'. There are good reasons for this: when putting a new kidney in there's no reason to take the 'old' ones out. The iliac fossa is relatively easy to access, and there are wide arteries and veins for the new kidney to be plumbed into.

The surgeon had opened a hole in Hennick's iliac fossa just above those iliac vessels. They had been dissected free of the tissues, raised up in loops, and closed off with metal clamps. One of the nurses opened the polystyrene box and I looked into it with astonishment; the kidney was cold, shrunken and a dusky grey – barely recognisable as an organ. It was lifted out and laid snugly into the hole in Hennick's abdomen. An assistant, one of the department's senior registrars, dripped ice-cold solution into the cavity to prevent its tissues warming to body temperature.

Hennick's iliac artery and vein, as well as the artery and vein of the new kidney, were spliced together with neat embroidery stitches. Then the surgeon took a deep breath, stretched his arms like a stage conjurer, and said to me: 'You're about to witness the most wonderful sight in the history of medicine.'

He removed the arterial and venous clamps in sequence, and Hennick's blood began to pump into the withered kidney. Each beat of his heart, visible in the pumping of the arteries, caused the kidney to swell. It was like watching a process of reanimation: a refutation of death. As the kidney grew, its defeated, dimpled surface began to fill out to a lucent pink. The surgeon held up the ureter of the new kidney (the tube which carries urine to the bladder) and I watched as a bead of urine began to grow at its cut end. 'It's working,' he said with triumph. 'Now we can stitch it into the bladder.'

Hennick's bladder had been filled up with antibiotic

solution by way of a catheter, and its outer surface stripped of fat. A tunnel through its outer tissues was made, about an inch long, and the ureter threaded through. At the far end of the tunnel a hole was cut into the bladder, and the ureter stitched into the free end. The surgeon placed a clear plastic drainage tube into the scar he'd made in Hennick's abdomen, then closed up his muscles and skin.

The operation was over: Hennick would be free of dialysis for life, although he'd rely on powerful drugs to prevent rejection of the new kidney by his own immune system.

A SUCCESSFUL KIDNEY TRANSPLANTATION is a triumph and a celebration, but is often achieved through exploitation of a tragedy. Until recently, kidneys for transplant have been obtained largely from the dead. Being involved in a successful transplantation is bittersweet; the relief of a life that has been saved is balanced by regret over the life that has been lost. I remember one that worked out successfully for several recipients, though catastrophically for the donor.

It was a night shift, 3 a.m., in a provincial emergency department. Paramedics were on their way with an unconscious teenage girl suffering a severe asthma attack. They had passed a tube into her windpipe to help her to breathe, but even with its help they couldn't get air to move freely through her lungs. When she arrived she was blue, and her mother and father were quickly ushered to the adjacent relatives room. We'd work to save their daughter with just a thin partition wall separating us. Anaesthetic gases can often relax the lungs, but with her they made no difference. We tried infusing drugs to widen her airways; tubes of high-flow oxygen; paralysing her muscles – but everything failed. Within minutes her heart began to beat erratically. All the clinicians were frantic, unable to accept that such a young woman might be about to die. We moved around her in a blur, glancing hurriedly up to a screen where her heartbeats began to broaden, then finally weaken.

She lost her pulse. My recollection of the next thirty minutes is hazy: adrenaline injections, chest compressions, atropine to quicken the heart's muscle. Twice her heart went into spasms of chaotic electrical activity and had to be shocked with the defibrillator, and after the second of these, her pulse restarted. Jubilation was followed by an evolving sense of horror: her heart might have restarted but her pupils no longer responded to light. She had regained a pulse, but had suffered severe brain damage. I phoned the closest city hospital, and its intensive care team made arrangements to come and get her.

Her parents were young themselves; must have been almost teenagers when she'd been born. I sat down, grey-faced, and explained with as much tact and as much truth as I could that her heart had stopped, had been restarted, but that her brain was no longer working properly. I told them she'd be transferred to intensive care, and they could travel with her. I can't remember the details of what I said, but when her father managed to reply, the spontaneous and transcendent generosity of what he said astonished me: 'If she doesn't come back, do you think she can help others?' he asked. 'Do you think she could donate her kidneys?'

She didn't recover in the intensive care unit, and after twenty-four hours or so, she was 'transplanted'. Her kidneys went to two different adults, at opposite ends of the country. Her corneas gave sight to someone who'd been blinded. Her liver went to a reformed alcoholic. Her pancreas and small intestine went to a teenage boy who suffered a rare genetic condition that meant he couldn't absorb food. Of her major organs only her heart and lungs – which had brought her to the gateway of death – and her brain – which had travelled too far into darkness to make it back to the light – were buried with her.

KIDNEY TRANSPLANT IS UNIQUE in that, because we have two, a single kidney can be donated in life with only relatively minor inconvenience on behalf of the donor. In the past these transfers of kidneys were for the most part between siblings, parents and children, but this need no longer be the case. Advances in tissue typing can match compatible organs across huge populations, and the acceptance of transplantation as a social good has meant more donations between individuals who are not blood relatives. These 'live unrelated donors' now constitute around half of all kidney transplant operations in the West, and occur between strangers. Since 2011 in the UK there has been a system of 'pooled donation' whereby someone can donate a kidney to an unrelated and unknown individual, and then others can donate in a gift circle that can be as wide as there are participants that can be lined up. Computers match compatible individuals.

B might want to give his kidney to his wife C, but as she's incompatible with him, she needs to receive one from A. Because his wife is receiving a kidney, B can choose to donate his instead to E. E's sister (D) donates one to G, and G's mother (F) donates one to H, and so on. It just takes one altruistic donor to start off the gift circle – in this case A – someone who donates a kidney to a stranger with no expectation of benefit to themselves.

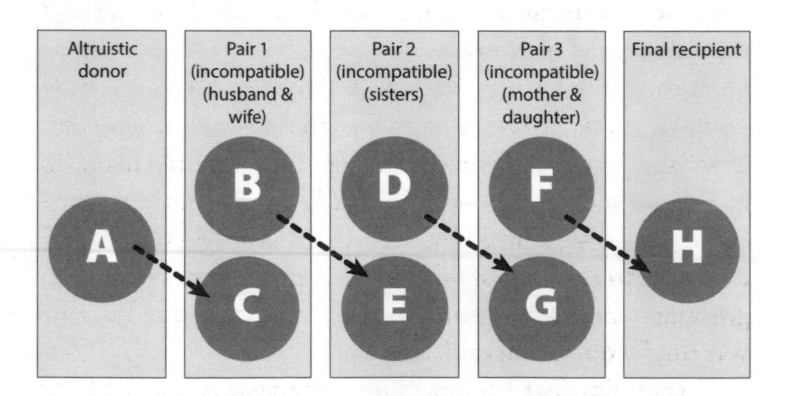

DAVID MCDOWALL is part of this new trend in sourcing kidneys for transplant in the West – the gift circle that can be started through an altruistic donation. We met through mutual friends at a time when he was recovering from the surgery. 'I was simply trading in a spare body part someone else can make good use of,' he told me. 'It wasn't much of an inconvenience for me, but could be a lifesaver for someone else.'

David never met the person who now carries his kidney, and because of the strict legislation around organ donation in the UK he never will. 'The risk of going through the operation was tiny, and besides, what is the point of a risk-free life?' David is a scholar and historian now in his sixties, who specialises in the Middle East. 'I have had far closer brushes with death working in Lebanon,' he said.

David had been thinking about donating one of his kidneys ever since he read an article in a newspaper about the possibility of making such a gift. Several years earlier he'd been close to death with a bleeding stomach ulcer, and would have died without transfusions. Donation was for him a fitting way to return a gift to the system that has saved his own life (blood transfusions are unpaid in the UK – historically a far more usual way of donating body tissue). When his grandson was born with a life-threatening condition requiring surgery, six weeks of intensive care and several months of hospital recovery afterwards, he felt the push to commit to going ahead. 'By then I knew I must do it,' he said. 'It was a sort of thanksgiving – even if my grandson had died I would still have done it, as the decision to donate had already been made. Having said that, I was acutely conscious of all that the health services had done.' He wrote to Hammersmith Hospital in London, offering to give them one of his kidneys, and just over a year later he was on the operating table.

I told him that I'd heard of some people, particularly if they'd been paid for their kidneys, being unhappy later about the decision – they'd found the experience more frightening and more painful than they'd anticipated. 'That wasn't my experience at all,' he said, 'the main difficulty, at first, was simply turning over in bed because of the discomfort of the scar, but that passed very quickly.' He'd had the operation at 9 a.m., and by that same evening had taken his first steps out of bed. 'Some intelligent doctor explained that the sooner I was walking the sooner I'd be out of hospital,' he said, 'so the following day I walked and walked and walked, clutching the stand with the drip. I was moved to an ordinary ward – had a very sleepless night – and they let me go the following day.' He'd been in hospital just over forty-eight hours.

'Are you curious about who has your kidney now?' I asked him.

'Of course!' he said, 'But I understand why they mustn't tell me. I'd hate for anyone to feel uncomfortable, or under any sort of obligation.' He became thoughtful. 'When I'm walking the city streets the knowledge that I may be passing someone carrying it, that I could even meet him or her and never know, is a delight.'

IN EUROPE IT'S COMMON to place stones on high ground in acts of commemoration, but in Tibet the hilltop memorials are more explicit. The traditional method of body disposal there is 'sky burial': the bodies of the dead are broken into pieces and left on a mountainside for vultures. It's a convenient way of disposal where the soil is too thin to dig graves, and a way of acknowledging that it's only through the deaths of living things that other lives are sustained. The soil around sky burial sites is littered with human bones, which remind travellers of the impermanence of all things.

Just as Europeans create cairns to guide travellers, Tibetans build piles of stones along traditional pilgrimage routes. These routes are like meridians over the landscape;

as pilgrims move along the paths they carry stones from one pile to another. As the circling of special stones over the sick can be seen in Tibetan medicine as a way to a kind of healing of the body, so the circling of stones over the landscape, carried in the hands and pockets of pilgrims, can be seen as a way to healing of the spirit.

Healing stones aren't exclusive to Tibet: in the town of Killin in Scotland there are a collection of eight stones held sacred to St Fillan, a Celtic divine thought to have been active in the eighth century. The tradition holds that you take the stone that most resembles your own afflicted organ, and rub it on your body. Visitors can go to the old mill in Killin, the first of which is said to have been established by the saint, and take the stones in hand. One looks like a face, one is marked like ribs, and another has an umbilicus like a belly. There's one that is dark and particularly smooth, and resembles a human kidney.

The poet and artist Alec Finlay has an interest in these sacred stones, and has combined it with his fascination with transplant surgery. He was commissioned by the government in Scotland to create a national memorial 'for organ and tissue donors', in the Royal Botanic Garden in Edinburgh. He built a traditional *taigh,* a Gaelic turf-roofed house, such as those found in the high country of Scotland – constructions which once offered shelter for pilgrims, herders and hermits. Visiting it, I was reminded of the Buddhist cairns and mountainous uplands of Tibet. *Taighs* weren't always built for shelter: there are some that were built for ritual and to house sacred stones.

'I felt that the memorial needed to make manifest qualities of inwardness and shelter,' Finlay wrote. 'I wanted there to be some kind of a protective dwelling for the feelings of those who were grieving … a chamber for the memory of the dead, but, being in a garden, it could gather a sense of floral growth and light.'

In the roof of his *taigh* Finlay laid a series of stones,

inspired by those in Killin, representing gifts of organs from the dead to the living, but also donations from the living so that others' lives were eased. On the floor of the structure there was a hollow cut into stone, as smooth and concave as a baptismal font, and around it a simple, nine-word poem engraved in a ring, repeating itself endlessly:

Finlay's intention was to celebrate remembrance and sanctity, and explore ways in which the body and its memories can be embedded in a landscape. But he also wanted the memorial to acknowledge that transplant is a new phenomenon, rendered possible only through high-tech advances in medical science: 'There is no curative treatment that is closer to being a secular miracle,' he said of transplant surgery – a miracle wrought through medical and surgical expertise rather than faith in the healing power of stones. Into the roof of the *taigh*, above ground, he had placed stones symbolic of transplanted organs, but under the *taigh* he buried a wooden chest, representative of the dead that have become donors, to acknowledge that what is most meaningful is often out of sight. Into the lid of the buried chest he secured a surgeon's scalpel and a packet of the medication used to prevent rejection of the transplanted organ.

To preserve anonymity, and emphasise how much we hold in common, Finlay hand-wrote the first names of every organ donor in Scotland in a book, interlinking each name to the others through a series of woven poems. The memorial in the botanical garden acknowledged the physical landscape around us – mountains and forests, cairns and sky burials – but also the social landscape of human connections to which we are bound.

12

Liver: A Fairy-Tale Ending

At last she called a Huntsman, and said: 'Take the child out into the wood; I will not set eyes on her again; you must kill her and bring me her lungs and liver.'

Little Snow White, *Grimms' Fairy Tales*

BLOOD-TEST RESULTS tend to come by computer these days, but when I was starting out in hospital medicine they came up twice a day from the mail-room on sheaves of pink, yellow and green paper. One of my jobs was to look through the papers then sign them to acknowledge receipt. If the results showed that a change in antibiotic was needed, or that a patient's kidneys had failed, it was my responsibility – as the one who'd signed the papers – to do something about it.

Pink was for haematology, listing the concentration, maturity and haemoglobin levels of the cells in each patient's blood. Yellow was for microbiology, detailing every virus or bacterium the lab had managed to isolate. Green was biochemistry, and listed those substances that give an indication of liver, thyroid and kidney function, as well as body salt levels. Each was tabulated vertically on a grid, beside older results so that trends over the course of days could be discerned.

			Blood	Blood	Blood	Blood	Blood
Urea	mmol/L	2.5-6.6	4.4	4.8	4.2	5.0	
Creatinine	umol/L	60-120	70	76	74	72	
eGFR (/1.73m2)	ml/min				72		
eGFR (/1.73m2)	ml/min					>60	
Sodium	mmol/L	135-145	139	141	141	139	
Potassium	mmol/L	3.6-5.0	3.6	3.5	3.9	3.2	
TCO2	mmol/L	22-30	27	25	25	25	
Glucose	mmol/L				5.0		5.8
Glucose spec.	type				FASTED		RANDOM
Bilirubin	umol/L	3-16		6	9	8	
ALT	U/L	10-50		22	22	22	
Alk.Phos	U/L	40-125		73	83	86	
GGT	U/L	5-35		40	40	37	
Albumin	g/L	35-50		42	43	40	
Cholesterol	mmol/L			5.8	6.0	5.0	
Triglyceride	mmol/L	0.8-2.1		2.2	2.6	3.0	

'Liver function tests' or 'LFTs' could be the most difficult to interpret, and are misnamed – they don't indicate much about the function of the liver. Instead they measure substances that are ordinarily contained within liver tissue, but which leak out into the blood in proportion to how irritated or inflamed the organ is. It would be more accurate to call them 'liver inflammation tests'. One of these, 'gamma glutamyltransferase' or GGT, rises in particular when the liver is inflamed by alcohol or gallstones. Another is 'alanine transaminase' or ALT – more likely to rise in hepatitis or when drugs or the immune system attack liver tissue. The liver is a mysterious organ: essential to life, multifarious in its actions, its tissue unusual in being able to regenerate. It detoxifies the blood and discharges unwanted chemicals into the bile. Another of its many functions is to create proteins that the body needs: a measure of this is the level of 'albumin' in the blood. Albumin shows how well the liver is managing to generate proteins, but also of how well nourished an individual is. If someone is starving, or their liver is failing, albumin levels begin to fall.

NIAMH WHITEHOUSE was in her late twenties, a small, neat woman with ink-black hair and pointed, puckish ears. I heard the story of her life, and her illness, from one of her work colleagues. She grew up in Edinburgh, an only child, and her father died when she was seven years old. At the age of fourteen her mother remarried and Niamh ran away from home – she subsequently lost all contact with her family. She had always loved to be outdoors, and after a few years drifting in London came back up to Scotland. She found a job as a junior gardener at a stately home and worked there happily for several years, rarely leaving the estate.

While digging among rose beds one day, she scratched her hand on a thorn. The wound bled, but she thought little of it. The morning after the rose scratch she didn't feel well – dizzy and unsteady, with a fever and aching muscles. She had to stop work early, and staggered back to her cottage. She wondered if she had 'flu. When the head groundsman arrived the following day to allocate her work she could barely make it to the door. 'Just stay in bed today,' he told her. Later, he peered through the window and saw her slumped on the sofa. She didn't answer when he rattled the windowpane so he broke down the door and called an ambulance.

I met her on the intensive care unit, paralysed and on a ventilator, with plastic tubes going into her nose, mouth, neck, wrist, forearm and bladder. Her eyes were taped shut to protect her corneas, and she had wires on her chest to record each heartbeat. A plastic clip shone red light through the skin of her earlobe – it provided a continuous readout of the oxygen level of her blood. She lay amid a forest of drip-stands seeping a cocktail of antibiotics, plasma substitute, transfusions and heart-strengthening drugs. Her hair straggled across the pillow like a black halo. During the struggle to put needles into her, some crimson spots of spilled blood had dripped down her neck and onto the hospital sheets.

Bacteria on the rose thorn called *staphylococcus* had entered her bloodstream and begun multiplying. Toxins spilling out from the bacteria were laying waste to the normal, harmonious control of her body's functions. Soon after she'd collapsed her blood became incapable of regulating how and when to clot: scarlet patches of haemorrhage bloomed across the skin of her trunk and limbs while other parts of her bloodstream began to clot off and starve her tissues of oxygen. Small clusters of bacterial growth began flying off into her fingers and toes, causing blackened smudges on her fingertips like the blight that browns the tips of leaves. Blood pressure is ordinarily maintained by tight seals along the lining of our arteries and veins, but chemicals produced by the conflict between her own immune system and the bacteria began to break down that seal. As a consequence her capillaries became leaky: her slim figure became as waterlogged with tissue fluid as a riverbank in flood.

At first the infection proliferated only within her bloodstream, but then some imbalance sent it whirling out into her other organs. Messenger proteins from her immune system began to confuse their targets, and liver cells became caught in the crossfire. I watched the progress of this collateral damage on the green biochemistry sheets. Albumin began to fall; as the cells in her blood broke apart the haemoglobin within them was metabolised to a waste product: 'bilirubin'. Her failing liver couldn't process the bilirubin into bile, or discharge it the usual way into the gall bladder, so the concentration of it within the blood began to rise. The bilirubin yellowed and stiffened her skin with jaundice, as if her body was embalming itself from the inside. Her GGT and then ALT began to rise; first to double the normal limits, then quadruple and beyond.

Twice a day on the ward round I'd gather with my seniors to peer at the grids of figures, trying to predict her path of recovery, or gather some hope from their trends. As she lay in bed it seemed as if she was in a state of suspended animation, but in truth every day brought her closer to death.

BEFORE IT WAS KNOWN that the heart was a pump, it was widely believed that blood was created in the liver, and flowed from there to the heart in a torrent driven by the force of its own generation. In the heart it was blended with vital spirit from the lungs then dispersed out to the tissues where it was consumed. As the source of blood, and thus of life, the liver was a symbol of power and mystery – examining it was thought to yield secrets about the future. It's a vast, solid organ, the biggest of the abdominal viscera, with wide-bore connections into the heart's ventricles and the intestinal system – no wonder it was thought to hold to secret to life. For Shakespeare, the quantity of blood in your liver said something about the strength of life within you: 'if he were opened, and you find so much blood in his liver as will clog the foot of a flea, I'll eat the rest of the anatomy'.

As far back as ancient Babylonia the livers of sacrificed animals were examined for their power to predict events. This method of divination is well documented in the Bible: the book of Ezekiel describes a king planning his next move by it. A priest who predicted the future from examining the liver was known as a haruspex: 'For the king of Babylon stood at the parting of the way, at the head of the two ways, to use divination: he made his arrows bright, he consulted with images, he looked in the liver.'

Another Near Eastern myth, that of Prometheus, seems to have recognised that the liver is the only whole organ capable of regeneration. Prometheus' punishment for having stolen fire from the gods was to be chained to a rock and have eagles tear at his liver, thus attacking the origin of his life. Every day it regrew, prolonging his torture.

The practice of divining the future from the liver wasn't confined to Mediterranean and Near Eastern cultures: the Roman historian Tacitus wrote in his *Annals* of the way northern Europeans sacrificed human beings, sometimes examining the 'palpitating entrails' to predict the future, – they weren't averse to eating them either. Even today 'I'd like to eat your liver' is a term of endearment from the eastern reaches of Iran as far west as the plains of Hungary. There may be echoes of cannibalism in the speech of Iran and Hungary, but in northern Europe, the tradition that Tacitus reported on has largely gone from the vernacular. It didn't disappear entirely, however: in the folk tales gathered by Jacob and Wilhelm Grimm there are echoes of eating the liver, as well as of using entrails to predict the future.

IN THE STORY OF LITTLE SNOW WHITE, the first version of which was published by the Grimm brothers in 1812,* it's not the examination of entrails that grants supernatural knowledge but a magic mirror – reminiscent of the Babylonian king's anxious consultation with 'images'. In the earliest versions, Snow White is just seven years old when her beauty surpasses that of her mother the Queen. 'Whenever she looked at Snow White,' the story goes, 'her heart heaved in her breast, she hated the girl so much.' She orders a hunter to take her daughter out and kill her, bringing back entrails – the lungs and liver – as proof of the murder.

It's curious that the liver and lungs were chosen as evidence, rather than the girl's head, or heart, or even her corpse. I asked Marina Warner, a distinguished academic who works on myth and fairy tale, why she thought entrails and in particular the liver were chosen in the original version of the Snow White story. 'Entrails give signs,' she told me, 'and the witch-like aspects of the wicked queen are enhanced by her closeness to a pagan haruspex in this, perhaps.' The huntsman couldn't bear to kill Snow White, of course, so presented instead the organs of a pig. According to the original Grimms' tale the Queen inspected them, was satisfied, then ate them 'salted and cooked'. With a

* The first version in 1812 was intended for a largely academic audience. For the second edition some tales were sanitised (for example, cannibal 'mother' replaced by 'stepmother') and explicit references to sexuality and pregnancy were removed.

better knowledge of comparative anatomy, or even of animal butchery, the Queen would have known she'd been tricked – pig livers are lumpier than ours, the lobes of which are relatively smooth.

When the wicked Queen finds out that Snow White is still alive (and living with seven dwarfs) she dresses as an old crone and delivers three poisonous gifts. The last of these is an apple, Eve's downfall: symbol of knowledge in the Genesis myth (and on the flip-top of personal computers). Snow White eats the poison and falls down comatose – almost as if she were suffering from blood poisoning.

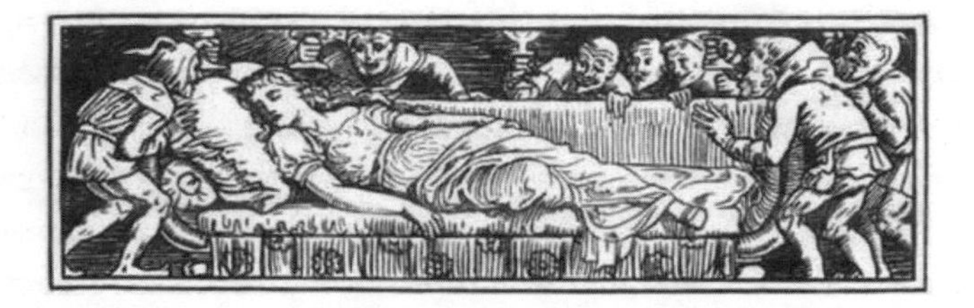

This time the dwarfs can't revive her, though 'she looked still as if she were living, with her beautiful blooming cheeks'. They put her in a glass casket in order that they could go on admiring her, and because it seemed a pity to bury such a beautiful, lifelike girl.

Snow White is one of many 'sleeping beauties': beautiful young women who fall asleep as if dead in European fairy tales and myths. The earliest sleeping beauty is found in a fourteenth-century French tale, *Perceforest*, and just as the original Snow White tale is far darker and more disquieting than the versions we've come to know, so is the original Sleeping Beauty, in which the girl is raped while in her coma, then gives birth without waking up. In a seventeenth-century Neapolitan version Sleeping Beauty gives birth to twins named for the Sun and Moon, one of whom rouses her by sucking a poisoned thread from her fingertip.

In Snow White, the girl's coma is broken not by the

traditional prince's kiss, but by dislodging a piece of the poisoned apple from her throat. It's as if the poisoning and her coma have been a time of adolescent transition; she throws off her glass casket and, like a butterfly breaking out of a chrysalis, emerges into womanhood and promptly agrees to marry the prince.

There's an enduring, puzzling fascination with stories of these passive, comatose, beautiful girls. The tales are heavy with symbolism about the maturation of sexuality, though the meanings ascribed to the girls' sleep seems to shift with time. They are continually being retold and updated for the new generation whether in film or animation. Marina Warner has written of how the Disneyfied retelling of these myths no longer have 'sweet, biddable girls: in family entertainment, heroines have become fast-talking, athletic and indomitable; they take on all-comers, especially would-be lovers, and they show no sign of falling in love'. These heroines may be dynamic, but an appetite persists to have them fall unconscious and emerge transformed from their sleep. In 2014 *Sleeping Beauty* was reworked by Disney as *Maleficent*: a dark, gothic fantasy in which an adolescent girl pricks her finger, falls comatose, and is awakened by a kiss of maternal, rather than matrimonial love – the redeeming kiss is given by the repentant dark fairy who issued the curse.

Recently, I watched Disney's take on Snow White again: *Snow White & The Seven Dwarfs*. When it came to the scene where the girl was laid in a casket of glass, it reminded me of an isolation cubicle on an intensive care unit.

NIAMH'S BOSS LOOKED through her cottage and found an old address book in one of the drawers. He began phoning around to find anyone who knew her family. After a few false starts he turned up an old school friend who gave a number for Niamh's mother. He phoned it, broke the news, and a couple of hours later she arrived at the hospital.

She was like a rococo cathedral: high, stately and with

an expensively dressed façade. Her voice tinkled like money. I explained as clearly as I could that Niamh had suffered septicaemia – blood poisoning – and that her liver and kidneys had partially failed. The crimson spots blooming over her skin were caused by the infection. Her heartbeat was weak, her liver was failing, and we were giving her transfusions and maximal doses of antibiotics. The mother's eyes were wide, scanning me, as if my face held the key to the future rather than just relating some details about the present. 'We don't know whether she'll survive,' I said, 'but the next few hours could be critical.'

'Well I shall stay right here,' she said.

The next set of biochemistry results showed little change, but for the first time there was no deterioration in her liver function. The following two mornings I arrived to find Niamh's mother asleep in a chair by the bed – it was as if she was making up for all the years that she and her daughter had been apart. I was more anxious than usual waiting for the following day's bloods to come back from the lab, and asked them to phone with the results: 'Good news,' said the lab technician. 'Her ALT is down, and her albumin has come up a little.' Another day, and there was further improvement in all the parameters we were measuring: the consultant thought we should try reducing her sedation. As we turned down the dose of anaesthetic her eyes began to move beneath their taped-down lids, as if she was trapped inside a dream world. The following day she woke up.

She woke up and saw her mother, and her smile was like an upside-down rainbow. Later that day she whispered her first words: 'I'd like to come home.'

NIAMH'S LIVER ALMOST FAILED – she came very close to dying because of her blood poisoning, and its consequent effect on her liver. But its tissue regenerated and brought her back to life. It was not some handsome prince that saved her, or any reconciliation with her mother – it was her own liver.

LFTs are among the most common tests I send to the lab; I look through grids of them every working day. Often they're raised because of alcohol – even slightly more than the recommended amount can double or triple the blood levels of GGT. Sometimes it's drugs: the statins that reduce cholesterol have a habit of sending liver tests awry. Gallstones block excretion of bilirubin, malnutrition drops the albumin, and sometimes a general inflammation revealed by the tests implies that cancer is darkly at work.

Occasionally, I can't find a reason for the liver's inflammation, so I send the patient to a modern-day haruspex for a biopsy. Through a keyhole in the abdomen the high priests of techno-medicine extract a piece of liver tissue, examine it carefully, then pronounce judgement on the patient's future. Even when their verdict is blcak, the liver can often regenerate; there's always the chance of a fairytale ending.

13

Large Bowel & Rectum: A Magnificent Work of Art

Midway, his last resistance yielding, he allowed his bowels
to ease themselves quietly as he read ... Hope it's not
too big to bring on piles again. No, just right.

James Joyce, *Ulysses*

HUMANS COULD BE DESCRIBED as tube-like animals, our skeletons and organs as elaborations to support a length of gut. From that perspective we're not all that different from nematode worms, primitive organisms that seem to exist primarily to ingest and excrete. Food goes in one end, faeces out of the other, and nutrients and water are extracted. In nematodes, it takes just a fraction of a millimetre to accomplish, but in us it's between twenty and thirty feet. Our bowels are forced into loops and spirals in order to fit in the space they're allocated; they squirm and twist constantly as they squeeze food and faeces along. The rectum is the terminus of that tube, and isn't free to move around – it's stuck down to the back wall at the spine. Its name comes from the Latin meaning 'straight': as the bowel jinks its way out of the sigmoid colon it makes a straight run through the pelvis for the exit.

In terms of function, the rectum is really just a waiting room: a place for faeces to accumulate until it's convenient to let it out. Bowel habit comes to most as a birthright: morning or evening, regular or irregular, loose or firm, we grow accustomed to the way waste exits, and alarmed if its pattern begins to change. For the most part that's with good reason: doctors are interested in changes to bowel habit because they can signal deeper disturbance. Diarrhoea can be a sign of thyroid disease, constipation a warning of malignancy, and oily, floating faeces suggest that your pancreas has packed in.

Just as a great deal of information can be revealed about someone's state of health from asking about how often they open their bowels, there is a lot to be gleaned from checking the inside of the rectum itself. In men it's the main way of examining the prostate, which can be felt by a (gloved) finger through the thin anterior wall. In women the cervix lies in about the same place, and in some women, particularly if they've never had sex, it's more acceptable to check the cervix rectally rather than vaginally. If someone is passing blood, an examination is necessary to find out if the blood is coming from haemorrhoids, from a tear in the anal skin, or from a tumour – I've found several rectal cancers this way (the medical school aphorism goes: 'If you don't put your finger in it, you'll put your foot in it.')

Stand-up comedians might suggest that to have your large bowel inspected you should drop your trousers and bend over, but the best way is actually to lie on your side, up on a couch, and draw your knees towards your chest. It's always surprising how many people apologise or make an embarrassed joke as they get into the position: 'I hope you haven't just had breakfast'; 'I'm so sorry that you have to do this,' as if the rectum is so sordid that, as the examiner, I might feel repulsed. It's an understandable belief: we're taught from our earliest years that faeces are untouchable, and that the rectum and anus are dirty and disgusting.

For most doctors, disgust at suppurating wounds, prolapsed bowels, or gangrenous limbs is beside the point: they have to be examined, so their aesthetics are irrelevant. But though ugliness has little place in the consulting room, there is still room for beauty, in the dictionary's sense of 'calling forth admiration'. The intricacy and economy of human anatomy, both in health and in sickness, is often beautiful. And if imagining the harmony beneath the skin can be beautiful, medical images such as ultrasound scans are, too – think of those grainy, chiaroscuro scans given pride of place on the mantelpiece or on the first page in a baby's album. X-ray images have a particular ethereal beauty to them, whatever part of the body they represent; contemplating them is a reminder not just of the skeleton and our mortality, but a way of transforming perspective and imagining the body anew. Sometimes they are like portraits, but they can also resemble landscape paintings with contours, horizons and cloudscapes. There are parallels in nomenclature: in emergency departments I've often ordered 'skyline' views of the knee, or 'panoramic' views of the jawbone. That those images have clinical importance, useful in diagnosis and treatment, makes them more, rather than less, beautiful.

The sculptor Rodin said that there was no ugliness in art if that art offered some insight of truth, and the same could be said for the practice of medicine, and the images that it creates. Medically speaking the body is rarely ugly, and images of it can have an aesthetic that approaches art – even if those images are of … the rectum.

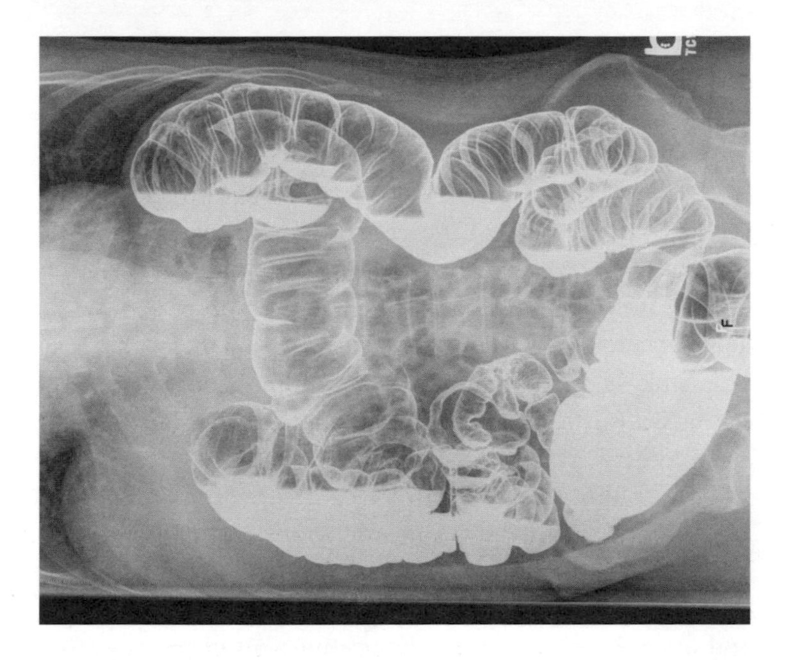

DOUGLAS DULETTO was a thin, middle-aged man who wore horn-rimmed spectacles and a starched white shirt. He had neat greying hair, parted in the centre, and sat primly on the emergency room gurney as if waiting patiently for the second half of a chamber music recital. He was wearing a thin hospital gown, and had neatly rolled up his corduroy trousers and placed them to one side of the gurney.

I picked up a clipboard from a holder on the cubicle wall and glanced down at the sheet: 'Foreign body, rectum' it said.

'I'm mortified to be here,' he said, flushing suddenly, 'but I can't get it out.'

'What's "it"?'

'A bottle,' he replied. 'I've been trying to get it out all evening.'

'A bottle of what?'

He flushed an even deeper scarlet: a senator snapped at a strip club.

'Ketchup.'

I asked him to lie on his left side with his knees drawn up to his chest – 'I've left my dignity at the door anyway' – then pushed a gloved finger into his rectum. 'Just bear down,' I said, 'squeeze as if you're trying to open your bowels.' At the tip of my finger, as far in as I could push, I could feel an edge of hard glass – too deep to get a finger on either side of it. I inserted a clear plastic tube – a proctoscope – and shone in a light. At the clear plastic edges of the instrument I could see healthy pink walls of rectum flecked with yellow sluglets of faeces. At the centre, just at the limit of my view, there was a glint of glass. 'It's going to be tricky I'm afraid,' I said, 'it's quite far in.'

He sank forward, head in his hands, and his shoulders began to shake. At the ward's 'sluice' – the area where all the urine and faeces are disposed of – I found a commode, and from the surgical ward some ointment ordinarily used in the treatment of tears in the anal skin. The ointment relaxes the sphincter, which can allow tears to heal, but I wondered if it would also allow the bottle to pass. I applied the ointment, and asked him to sit on the commode.

After he'd strained a few times I got him back up on the couch, then tried for the bottle again. This time I thought I had it, when at the last minute it slipped away deeper into the swampy anatomy of the abdomen. I swore under my breath, but he heard me.

'What's wrong?' he asked, nervously.

'Nothing,' I told him. 'But we're going to have to get an X-ray.'

At that time X-rays were still produced on large acetate films. Once Mr Duletto was back in the cubicle I took the envelope containing the film back to the doctor's room, and put it up on a light box. It gathered quite a crowd.

The bowl of the pelvis stood in the foreground, shaped like the two flanks of a valley, beneath vague, gaseous bowel shadows – a Turner-like sky. Rising up through the middle was an incongruous form: a skyscraper dropped into a pastoral scene. It was the crisp, instantly recognisable outline of a branded bottle of ketchup. It lay along part of the rectum and into the sigmoid colon, with the shoulders of the bottle and its metal lid tapered like an arrowhead pointing deeper into the guts.

'I'm sorry,' I said when I got back to the cubicle, 'I'm going to have to refer you to the surgeons. There's no way I'm going to be able to get that thing out on my own.'

ACCORDING TO THE PSYCHOLOGY of aesthetics, art is appreciated not just through the perception of something as beautiful, but because it can elicit a wide variety of emotions: confusion, surprise, disgust and even embarrassment. Looking at the X-ray, it undoubtedly had an aesthetic value: the grainy forms of bone and bowel against the moulded artistry of glass and metal. There was pop-art appeal to the juxtaposition of a mass-produced bottle against the organic shape of Mr Duletto's pelvis. This X-ray is a work of art, I thought to myself: it could be submitted to a gallery, or projected at night onto the hospital building. I pictured it for a moment hung in MoMA or the Tate Modern, protected by glass and cordoned by rope.

I dictated a letter for the surgeons, and a porter came down to take Mr Duletto up to their ward. 'Surgical?' the porter asked, and I pointed to the cubicle. He pulled the gurney out into the corridor, and Mr Duletto lifted a hand to wave as he headed towards the door. 'Any X-rays?' the porter called out.

'Oh yes,' I said, turning to the light box, but there was no X-ray there. It had been stolen: someone else must have appreciated it for a priceless work of art.

PELVIS

14

Genitalia: Of Making Babies

I wish either my father or my mother, or indeed both of them, as they were in duty both equally bound to it, had minded what they were about when they begot me.

Laurence Sterne, *Tristram Shandy*

TO CONTEMPLATE THE OBSTACLES to conception is to dig into the deepest ideas of what it means to be human. Did our lives begin when the ball of cells, of which we were once composed, first bounced against the wall of our mother's womb? For many women, fertilised eggs won't implant to the womb lining. Did it begin further back, when the fastest, strongest sperm from our father fused to an egg from our mother? Some men have sperm too sluggish or disorientated to find an egg. Was our life decided three months earlier, in the genetic dance called *meiosis*, when the successful sperm that created us was created deep inside one of our father's testes? In some men meiosis doesn't work properly; they are *azoospermic* – have no sperm in their semen. Or perhaps our individual selves were generated just two weeks earlier, when the egg that went into creating us won the privilege of being primed for ovulation. Disordered menstrual cycles, and failed ovulation, are a frequent cause of infertility. In a sense our lives began decades before our parents even came together – the eggs in our mother's ovaries were created when she herself was in the womb.

Then there are the physical obstacles to the egg reaching the womb: the fallopian tubes have tiny projections on their open ends which gather eggs the way fingers might gather jewels. When each of us was fertilised, our primordial, cellular selves began to divide high in the fallopian tube: one cell became two, two became four, four became eight, and so on. Like a city crowd around a royal pageant, cells within the wall of the fallopian tube pushed the dividing mass of cells towards the womb. By the time it arrived, the fertilised egg had become a ball of cells numbering sixty or more.

While still within reach of the ovary the egg can be fertilised too early and drift off into the wrong part of the abdomen. This is one of the surprises of our anatomy – men have no equivalent connection between the inner and outer worlds as women have, to conduct sperm from the vagina to the inner abdomen. If a fertilised embryo implants deep in the abdominal lining it may even grow for a while, but is doomed to miscarry because the lining can't provide enough blood for a developing baby. If it miscarries internally in this way, the woman may not even know she was pregnant; over time, the embryo's tissues are replaced by brittle, bone-white calcium salts. Surgeons sometimes find these *fetoliths*, or 'stone babies', inside the abdomens of elderly ladies, carried unknown for forty or fifty years.

Occasionally, the developing embryo embeds itself part way along the fallopian tube: the commoner type of 'ectopic' pregnancy – meaning pregnancy in the wrong place. As the growing baby takes up space the tube is unable to expand; the embryo is doomed and the stretching begins to cause terrible pain. If the pregnancy is allowed to go on the tube itself will split and the mother may bleed to death – a poisonous gift from new life to old.

UNTIL THE LATE EIGHTEENTH CENTURY, in Europe, it was

believed that for conception to occur it was as important for women to reach orgasm as it was for men. A textbook of midwifery in use in the seventeenth century declared that without a clitoris women 'would have no desire, nor delight, nor would they ever conceive'. Judges presiding over cases of rape would decide that if conception occurred, intercourse must have been consensual. As late as 1795, the Marquis de Sade – who was greatly preoccupied with methods of avoiding pregnancy – was able to write that fluid 'discharged' by women during climax was a prerequisite for the creation of new life: 'Of the commingling of these liquors is born the germ which produces now boys, now girls.'

Though many societies had realised that this can't be true (female circumcision, for example, virtually precludes it), these ideas about the body had been around for thousands of years: new life was generated through a convulsion that necessarily had to be experienced by both sexes for it to work. Orgasm in women was assumed necessary for ovulation, but simultaneous orgasm was even more likely to result in a pregnancy. In the Hippocratic treatise *The Seed* the author describes how heat is created within the pelvis of both men and women during sex, leading to a paroxysmal climax which would be experienced more intensely by women if occurring at the moment when semen also made contact with the cervix ('like a flame flaring when wine is sprinkled on it'). Galen wrote that backache and limbache was common among widows who no longer had sex because of a build-up of female generative fluids within; the cure was to encourage discharge of this fluid, preferably through sex but if necessary through manual stimulation. In the sixteenth century the Dutch physician Forestus advised women to engage a midwife to carry out this task 'so that she can massage the genitalia with one finger inside … and in this way the afflicted woman can be aroused to the paroxysm'. This perspective on female sexuality went on in attenuated form until the early twentieth century: vibrators were invented for the treatment of women suffering from 'hysteria', and their use recommended right up until the diagnosis itself was struck out of the psychiatry textbooks in the 1950s. (Some of these devices had fittings so that they could be driven by the home sewing machine.)

Fig. 35.
Mov. XXII.

ROB AND HELEN came to my clinic eighteen months after throwing away Helen's contraceptive pills. They were awkward with embarrassment as they took their seats. 'We've been trying for a baby for ages,' Rob began, then hesitated, but Helen finished the sentence: 'We're starting to think there's something wrong.' He was a chef: tall and slightly overweight, with silvering hair and anxious eyes. She was an assistant at a nursery: slim with bobbed red hair and doll-like, porcelain-white cheeks. 'I don't know if we need IVF?' asked Helen, spinning her wedding ring with the fingers of her right hand, 'but at thirty-seven I'm told we should hurry up.'

I asked about family history. Helen was one of three children, didn't know of any problems that ran in her family, and both her brother and her sister already had children of their own. Rob was also one of three: though his brother had a daughter, she'd been conceived with the help of IVF.

On average, couples who have regular unprotected sex have about a 20 per cent chance of conceiving within a month, 70 per cent chance of conceiving within six months, and 85 per cent chance of conceiving within a year. It's for that reason that doctors prefer to wait at least a year before initiating infertility tests. The first tests to be carried out are the most straightforward: for Rob, two semen samples sent in at least a month apart after a few days' abstinence, and for Helen, blood tests at two separate points in her menstrual cycle to assess whether she was ovulating regularly. The semen samples are trickiest to arrange; they have to be delivered to the lab, which is only open at certain hours, within an hour of ejaculation. 'What ... these?' said Rob when I handed him the specimen tubes. 'They don't give you much ... to aim for.' How he went about obtaining the samples we left undiscussed. Helen laughed, dissolving the tension in the room at last. 'What are you trying to say about your equipment?' she said, elbowing him.

Helen needed a blood test on the third or fourth day after her next period began, followed by another one seven days before the *following* period was due. The first test gives an idea of whether the two hormones that coordinate ovulation – 'luteinising hormone' and 'follicle-stimulating hormone' – stand in the right ratios to one another and to levels of oestrogen. The second test gives an idea of whether the ovary is creating enough progesterone – the hormone that prepares the womb for pregnancy – to suggest she had ovulated. Helen drew her diary from her bag, where all her periods for the past year had been plotted out on a grid. 'This is my menstrual map,' she said grimly; 'a map of disappointment.' We picked out the days she'd need blood tests, and fixed the appointments.

When I met her next she came alone. After taking the blood samples she rolled down her sleeve and paused. 'You know the worst of it?' she said. 'It's what it's done to our sex life ... I mean, it's difficult to feel romantic, or desirable, when all you're thinking about is ovulation and conception.'

'Some people don't conceive until they get their appointment through for the fertility clinic,' I said, 'that's when they stop worrying about it. Don't make it a trial, or something to get stressed about.'

'That's just it,' she said. 'Before, I hardly ever had an orgasm with sex. Now, I never do. Do you think that's a problem?'

THE NERVE THAT COORDINATES ORGASM, called 'pudendal', has an almost identical course in men and women. Its name comes from the Latin, *pudere,* to be ashamed, as if we're still cowering in the Garden of Eden, trembling behind fig leaves. The pudenda might be comic, absurd or even embarrassing, but never shameful: without our parents' pudendal nerves, after all, few of us would be here. People can be reluctant or embarrassed to discuss aspects of conception, sex and sexuality, but as a doctor it's unavoidable; you can't work with human bodies for long without having to talk about them.

Whether folded in foreskin, or desensitised by circumcision, the pudendal nerve in men branches through the skin of the glans penis, and in women through the clitoris. Those nerves coalesce into bundles that run down the back of each *corpus cavernosum* – the 'cavernous bodies' present in both sexes, which stiffen through being engorged with blood, but which were once thought to be inflated by the *pneuma,* or spirit, of sexual desire. The nerve on each side then drops down into the penile or clitoral root and loops under the arch of the symphysis pubis of the pelvic bone – an angled Gothic arch in the male, and a rounder Roman arch in the female (with its smoother accommodations for a baby's head, and its more dissipated scatter of nerves). It then tunnels deeper into the layers of muscle and sinew that support and give continence to the bladder, taking in out-branches that supply sensitivity to the skin between the thighs. It's here that it slips under the prostate gland and seminal vesicles in men, which store and bathe the sperm that have migrated up from the testis, and the cervix and womb in women. Then it continues towards the spine, emerging into the pelvis between powerful muscles that cantilever the weight of the body into the legs.

The sacrum is a triangular bone at the base of the spine, perforated by holes like a priest's censer. It is so called because it was once believed to be sacred: a reservoir of human essence – medieval Europeans thought that at resurrection their bodies would be reconstituted first from the sacrum, and that energies discharged from the sacrum were essential in the creation of new life. After twisting themselves through the tangle of the sacral plexus, pudendal nerve fibres slip through the sacrum's perforations, and plug into the spinal cord.

Marcus Aurelius spoke of orgasms as the simple product of a timed duration of friction. Aristotle thought that the heat necessary for conception was generated by sex just as a fire can be ignited by rubbing two sticks. But of course the propagation of sexual tension is less predictable than those theories suggest; less a process of ignition than the interplay between storm clouds and an ionising earth – the lightning flash of a two-way traffic between the mind and bodily physiology. In Western countries where surveys have been attempted it's been reported that only a third of women regularly experience orgasm during intercourse, the reasons for this being both social and physical. The effect of drugs can play a part: antidepressants like Prozac and Seroxat, some of the most commonly prescribed drugs in the Western world, can so dampen the action of those nerve endings that orgasm becomes difficult to achieve for both men and women. Heroin can do the same, and, most famously, so can alcohol.

A mirroring tension builds between the nerves within the glans or clitoris and the answering plexus in the pelvis until some final, pivotal change provokes the climax. What the French called *la petite mort* can be seen on brain scans not as a darkening to oblivion, but as a 'lighting up' in the emotional core (cingulate gyrus), reward centres (nucleus accumbens), and hormonal regions (hypothalamus) of the brain. It's those hormonal regions that in some animals actually provoke ovulation as a response to sex, just as Galen imagined, but in humans that's not the case.

During orgasm, pulses of nerve stimulation ripple back out from the spinal cord to the prostate gland and seminal vesicles in men, and the cervix and vagina in women. In men they trigger the prostate, vas deferens and urethra to squeeze sperm and seminal fluid towards the penis in a series of clenching spasms, while coordinated reflexes shut the entrance to the bladder so that semen can go only one way – out. In women, those same ripples trigger convulsions in tiny glands around the urethra and anterior wall of the vagina – Skene's glands – which push out a sort of female seminal fluid similar to the prostatic fluid expelled by men.

The outlets of Skene's glands vary between women: on climax they may push a watery fluid out into the urethra as occurs in men, or directly into openings within the vagina – explaining why some women feel as if they 'ejaculate' on orgasm, while others do not. An Italian sexologist, Dr Emmanuele Jannini of L'Aquila University, believes that the area around the urethra on the anterior vaginal wall is a separate erogenous zone in some women, distinct from the clitoris. Like Ernst Gräfenberg, the New York sexologist whose initial 'G' gave the name to the 'G-spot', Jannini thinks that there are women who experience orgasms deeper in the vagina than others, as an accident of their pudendal nerve anatomy.

The vagina in health is acidic, something that helps keep it free of infection. Unfortunately, sperm prefer a neutral environment – neither acid nor alkaline – similar to that prevailing within the womb. The secretions from Skene's glands and the prostate gland are alkaline, which suggests that they helpfully neutralise the acid environment of the vagina at the moment when sperm are released into it. The secretions from Bartholin's glands, which lie at the posterior entrance to the vagina and become active much earlier in intercourse, are also alkaline and so do the same thing.

William Taylor wrote over two centuries ago, 'so the poetic orgasm, when excited, glows but for a time': in men, up to ten seconds; for women, orgasm can last double that. The pattern of female orgasm is different from that of the male: broader and slower to rise as well as fall away. There are several theories, none entirely convincing, which suggest how female orgasm might help in conception.* One theory is that the longer duration of the female orgasm in women could give the cervix more time to pull in male seminal fluid, which may increase the likelihood of pregnancy, and could help sperm survive by neutralising the natural acidity of the vagina. But there are others: by encouraging more sex; by secreting the hormone oxytocin from the brain (which may cause the womb to draw in fluid); even that female orgasms help in sexual selection – identifying men who are more likely to prioritise their women's happiness as highly as their own.

FOR SIGMUND FREUD, 'Eros' and the erotic represented the sexual components of life: churning with energy, chaotic and generative. He opposed it to humanity's drive towards aggression and self-destruction – the Greeks would have

* There's even a theory proposing that orgasm helps select sperm from a non-regular partner over that of a regular one. See R. R. Baker and M. A. Bellis, 'Human sperm competition: ejaculate manipulation by females and a function for the female orgasm', *Animal Behaviour* 46(5) (1993), 887–909.

called it 'Thanatos'. Carl Jung thought the erotic was less about opposing violence, and more about achieving a balance between the rational and the emotional aspects of human nature. 'Woman's psychology is founded on the principle of Eros, the great binder and loosener,' he wrote, 'whereas from ancient times the ruling principle ascribed to man is Logos' – from logos comes our idea of logic. For Jung, just as acids and alkalis have to balance to create a neutral environment, so the logical and the erotic had to balance for both men and women to flourish. In counselling infertile couples, Jung might well have characterised a dependency on blood tests and analyses, such as those taking place in an infertility clinic, as too great a focus on Logos, but a concentration solely on the emotional and sexual health of a couple's relationship as an overindulgence of Eros.

A few weeks later I met Helen and Rob again. Rob's semen analysis was normal: I ran through the parameters examined by the laboratory, translating the arid terminology of 'motility', 'morphology', 'concentration' and 'consistency'. Helen's hormone tests too had come back as I'd hoped: the LH and FSH were in appropriate proportion to one another, the oestrogen as low as it should be early in the cycle. The progesterone level in her blood a week before her period was due suggested that she'd ovulated normally – there was no obvious reason they weren't conceiving.

'So the results are all very reassuring,' I told them. 'Rob, your tests are normal, and Helen, your ovaries are ovulating at the time in the month we'd expect them to.'

'So what could be wrong?' she asked.

'Sometimes the tubes inside aren't letting the sperm past for some reason, sometimes the immune system prevents the sperm and the egg coming together, often there's nothing wrong at all.'

'So what now?'

'Now I write to the fertility clinic at the hospital, and you two try not to worry about it too much.'

WHEN THEY CAME BACK to see me a few months later their initial embarrassment had been replaced by dejection.

'How did you get on at the clinic?' I asked.

'Don't ask,' Rob said.

At their first clinic appointment Helen had admitted to the odd glass of wine, and was told baldly that she must swear off alcohol. Rob was irritated by the suggestion that he should lose a little weight, and by the in-depth questioning about how often, and in what manner, they had sex together. 'I suppose they have to ask,' he said, 'but it was as if they thought we didn't know where babies came from.'

After a further blood test, and an ultrasound scan of her ovaries, Helen was told she had 'depleted ovarian reserve'; there were relatively few 'follicles' in the ovaries with the potential to ovulate. The couple were likely to need IVF, but even at that their chances of success were small, at around one in ten. 'And you didn't warn me about the ultrasound,' she told me. 'I got a shock when the doctor rolled a condom over a plastic truncheon, and told me where he had to put it.'

Despite the indignities of the clinic they decided to go ahead. The first step in treatment was a series of injections to 'reset' all the follicles in Helen's ovaries, so that all were at the same early stage of development. Then she began a further series of injections, this time to hyperstimulate egg maturation – the development of many eggs at once. 'I couldn't stand those injections,' Helen told me. 'My bum was black and blue from them.' The internal ultrasound scans were by now so frequent they no longer bothered her.

Helen's ovaries began to swell with developing follicles, and a further injection provoked the final maturation of the eggs. Within thirty-four hours of receiving the injection, almost to the minute, the eggs were ready to be gathered. For that procedure she was given a powerful sedative and, using an intra-vaginal ultrasound scanner, a very fine needle was passed through the walls of her vagina and into her ovaries. The fluid within each follicle was drawn carefully out and examined for eggs. Rob had to provide a fresh sample of semen that same morning, then he and Helen were sent home.

That night Helen slept deeply thanks to the sedatives still soaking in her blood. Rob couldn't sleep for the thought that, as he and Helen lay together, his sperm and her eggs were being mixed together in a glass dish in some white-walled laboratory.

'They took the eggs on the Friday,' said Helen, 'and then on the Tuesday I had to go back. They had six fertilised embryos, two of which were of "good quality", whatever that means, and one of those – the one they said was the best – was put inside me.'

'And then?' I asked.

'And then it didn't work.' She looked away, and Rob reached across to take her hand. 'They told us the chances weren't good,' she said, 'now we'll just have to think about whether we can face it, or even afford it, again. They still have some of our embryos in a *freezer*. Maybe I'm frigid after all ... they'll feel right at home in there.'

FOR GALEN 'BARRENNESS' was the result of a lack of heat; to treat infertility the answer was simply to find ways of heating up the pelvic organs. This could be done with foreplay or 'lascivious talk', or by rubbing the genitals with herbs to redden and irritate the skin. Avicenna, the Arab physician of the eleventh century who transmitted much of this rhetoric back to the West, agreed that it was necessary to find ways to increase female sexual pleasure: '[when women] do not fulfil their desire ... the result is no generation', he wrote. At the same time, too *much* heat was thought counterproductive: prostitutes were considered to conceive only rarely because at that time it was believed they had too much ardour for sex, so that their seed was 'burned off' by excessive lust.

In *The Sicke Woman's Private Looking Glass* of 1636, John Sadler – one of the first English gynaecologists – wrote that the problem was often 'the man is quicke and the woman too slow, whereby there is not a concourse of both seeds at the same instant as the rules of conception require'. Rather than blaming women for infertility, Sadler laid responsibility on men to refine their 'allurements to venery ... that she may take fire and be enflamed'.

The assumption that women conceived in response to orgasm, which had been around for as long as we have written records, at last began to crumble when in 1843 a German physician, Theodor Bischoff, demonstrated that ovulation in dogs occurred even when there had been no intercourse. That same year a paper appeared in the medical journal *Lancet* asserting, wrongly, that the cycle of animals going into 'heat' was one to which 'the menstrual period in women bears a strict physiological resemblance'. Medical knowledge had woken up to the fact that women ovulate cyclically rather than as a response to sex, which not only fed into the new Victorian prudishness about female sexuality (if pleasure isn't necessary, why bother with it?) but gave rise to the mistaken belief that the fertile time of the month was during menstruation, which was the human analogue of animals 'going on heat'. It's a belief that persisted for nearly a century: in the 1920s Marie Stopes's bestselling *Married Love* advised that maximum fertility occurred just after the end of menstruation – more than ten days too soon. According to Stopes, it was mid-cycle that women were unlikely to conceive – exactly the time when we now know that pregnancy is most likely to occur.

A FEW MONTHS LATER Helen and Rob tried again, using the second of the two embryos that were deemed 'high quality'; but were again disappointed. 'Perhaps it seems crazy,' she said when she came to talk to me about the failure of the second treatment, 'but I want to have a baby so much; every time I pass a baby in the street, or pick one up, my womb does a spin. I don't know if I can go on working in a nursery.'

'Do you think you'll try a third time?' I asked.

'We can't,' she sighed. 'We've already spent all our savings on paying for that second round. By the time we've saved more, I'm sure it'll be too late.'

We were silent for a moment.

'And how are things between you and Rob?'

'Fine, actually; more than fine. It's a funny thing but …' She paused, as if wondering again how much intimacy to share. 'We're both upset about it, but in some ways we're closer than ever. What's that quote – "When you can't change the wind, adjust your sails." Things have been much, much better – for me as much as for him.' She blushed. 'Now that we've given up on trying to make a baby, it's as if we've been able to go back to making love.'

THERE ARE ASPECTS of our bodies' workings that even now, in the twenty-first century, remain obscure. It wasn't until the 1960s that the delicate hormonal weave between brain, pituitary gland and ovaries was unpicked with respect to fertility, and until the late 1970s that the first IVF baby was born. Despite all the advances of the subsequent decades, much remains hidden.

I've known women whose immune systems repeatedly mistook the embryo within their womb for an infection – and destroyed it. After suffering recurrent miscarriages they conceived only after suppressing their immune systems with chemotherapy-like drugs. I knew a couple whose recurrent miscarriages spanned a decade until, having called in a plumber for a burst pipe, were informed that they'd been drinking water corrupted by lead. When the ancient piping and cistern was removed they had no more problems. I've known couples each of whom was 'infertile' until they separated and found new partners – suddenly, both were able to conceive.

It was only a couple of months later that I saw Helen and Rob on my consulting list again. As I stood up to call them from the waiting room I wondered if they had changed their minds, and had found the money for a third course of IVF.

Usually when I stood at the waiting room door I'd see them nod, gather their bags, and solemnly get to their feet. But their manner this time was different: Helen's face shone as she looked up. We walked the few steps to my office, and she skipped the last couple to the door. 'You'll never guess,' she said before we had sat down, ' – I'm pregnant!' Without laboratories or relationship counsellors they'd found the right balance between Eros and Logos on their own.

15

Womb: Threshold of Life & Death

I see the elder hand pressing receiving supporting,
I recline by the sills of the exquisite flexible doors,
And mark the outlet, and mark the relief and escape.

Walt Whitman, *Song of Myself*

THE TV TOOK UP more space than the fireplace, but no one was watching it. A two-bar electric fire glowed in the dark socket behind the hearth. An ashtray shaped like a porcelain Pekingese dog was overflowing, and a confetti of cigarette butts littered the carpet. Along a line between the room's entrance and the patient's easy chair the carpet was worn thin; a trail greasy from the passage of dropped food and slippered feet. The sofa was longer than the room was wide, and seated on it were a man and a woman – the son and daughter of my patient. Both of them had to sit with knees splayed, to make room for the sag of their bellies. The son stood up to greet me, hands trembling.

'She's bleeding, doctor,' he said, 'from down below ...'

Parked outside in the car, before stepping out into the rain, I had read Harriet Stafford's medical history on the emergency service laptop. It read like a primer in the co-morbidities it's now possible to sustain through modern Western medicine, beginning with Emphysema, Coronary

Heart Disease, High Blood Pressure and Diabetes – the four horsemen of the ageing society's apocalypse. Beyond those usual four there were two other significant entries: 'Multi-infarct Dementia' explained her absent manner as she watched me approach, and 'Endometrial Carcinoma – Palliative' explained the bleeding – she was haemorrhaging from a cancer of the womb. At the end of the list was the plea, written by her own doctor: 'Avoid admission if possible.'

'Hello, I'm Dr Francis,' I said to her. 'How are you getting on?' Her eyes startled with the customary panic of the demented – afraid she'd answer wrongly, or make a fool of herself. I pictured the circuits of her brain, as worn by routine as her carpet. Instead of the expansive possibilities of social intercourse she was left with a few reflex replies. Some people with dementia return almost to a pre-verbal state; like very young children they learn to trust or distrust not through words, but through tone of voice and a speaker's manner.

'Nice, yes, fine,' she said, smiling up at me and dropping her guard a little. I picked up her hand and shook it gently. It was cool, her palm clammy, and her pulse was thin and rapid. 'I've come to help you,' I said. With the flat of my fingers I brushed the skin further up her arm; it was cold as far as her shoulder – she had lost so much blood that there was not enough left in her body to keep her limbs warm. The skin of her face was pale as candle wax, almost translucent. The whites of her eyes were bloodless.

'I changed her pad half an hour ago,' said her son. 'But the cancer ... it's gushing out of her.' He blushed at having to describe two taboos – cancer and vaginal bleeding – to a strange man.

'I'm going to have to examine her. Can we lie her down somewhere?' Off the hall was a small spare bedroom – she was no longer able to manage the stairs. Her son and daughter helped her up from the easy chair and, taking her arms as if encouraging a baby to walk, half supported and half carried her through. 'It's alright Mum, it's alright,' murmured the daughter, like a parent comforting a fretful child, before lifting her with ease and laying her down on the bed.

She lay flat on the bed, and I loosened her dressing gown. She had no idea who I was but the memory of doctors, and my appearance in a tie and white collar, echoed something within her and she accepted that being undressed like this was no cause for distress. Her blood pressure was so low it was almost unrecordable. 'Sore?' I asked her, trying to keep my language as simple as possible. She made a wincing expression and passed her hand back and forwards over her stretch marks. It seemed suddenly incredible that this son and daughter were once inside her womb; that her womb, having sponsored their lives, was now hastening her death. Pulling down her pyjama trousers I saw blood pooling in the pad, slick clots of crimson.

From a pack of vials in my suitcase I drew up some morphine, and injected it under the skin of her belly. The injection site was inches away from the tumour that was eating away at her womb, stiffening the organs of her abdomen and killing her as surely as if she had slit open her veins. As I stood watching her for a moment she closed her eyes and began to doze. On the wall above her head was a poster print of Jesus, with a bleeding heart and a Hollywood beard. Stacks of videocassettes were piled along the skirting boards. There was an open night bag, like the ones expectant mothers keep, stocked with talcum powder, cigarettes and spare nighties. 'We keep it there in case she has to go into hospital,' explained her son.

'Should we sit down next door and have a chat?'

They nodded, and together we moved back through to the living room leaving Mrs Stafford lying on her bed.

'I know you've not met me before, and I've just met your mum, but I can see on her records that she has a cancer, and we know she's bleeding from that cancer.'

'Aye,' said her daughter, nodding. 'They gave her weeks to live, and that was months ago.'

'Well, she's losing a lot of blood, and we could do one of two things. We could send her into hospital for a transfusion, or we could keep her here, and see what happens ...'

Her son and daughter looked at one another, until the son broke gaze and turned to look out the window.

'... and what might happen is that the bleeding stops, and she rallies, and things go back to the way they have been. Or what might happen is that she keeps on bleeding and fades away.'

'How long has she got?' the daughter asked.

'I wish I knew, but ...' I hesitated for a moment, then met her eyes, '... she could die tonight.'

'Just leave her here,' her daughter said, decisively.

'Alright,' I said, and a few moments passed. 'I'll come back in three or four hours to see how she's getting on.'

Before I left I wrote up notes in the district nurses' folder by the bedside, and helped her daughter change the pad. As I was pulling up her underwear I saw that the new one was already scarlet with fresh blood.

IT WAS THREE in the morning before I could come back. At the door I was met by her granddaughter who, in her haste to get to me, tripped and fell forwards, butting her head against the glass. 'The priest's in,' she gasped as she opened the door to me. She was heavily pregnant.

I stopped at the doorway, holding my case, wondering if my expression was serious and pious enough for a meeting with a priest over a deathbed. I felt a spasm of guilt that it was my warning – 'she could die tonight' – that had brought him out in this weather. There were ten people in the room including the priest: a tall, well-built man in his late forties or early fifties – better nourished as a child than his parishioners. He nodded to me from the foot of the bed. From my doorway glance, I could see that Mrs Stafford had already drunk the blood of Christ, taken the viaticum of last rites, and now lay propped up on pillows.

I waited just outside the doorway. On the sofa behind me I could see that the folder I had written in was lying open; the whole family had been poring over it, as if over tea leaves. The prayers went on for ten, fifteen minutes more. And then there was a bustle, and one by one Mrs Stafford's son and daughter, her granddaughter and several grandsons, began to leave the room. 'Evening, Father,' I said to the priest as he nudged past me on the way out of the room.

'Evening, Doctor,' he said, clapping me on the shoulder and giving me a quick, businesslike smile; 'it's fine work you're doing.' Before I could offer, 'You too,' he was already gone.

I entered the room; Mrs Stafford opened her eyes and I took her hand, wondering if she recognised me at all. 'I met you earlier,' I said, 'I'm the doctor.' She grunted an acknowledgement, closed her eyes again and laid her head back on the pillow. This time her pulse was faster, and I couldn't find her blood pressure at all. Her hands and feet were just as cool as they were earlier. 'She says she feels cold,' her daughter added, coming in from the living room behind me. 'We've got the electric blanket on, but ...'

I undid her dressing gown again and began to gently press on her belly. She uttered a low moan, and I drew up another vial of morphine, and again injected it into the skin of her abdomen. 'Have you had to change the pads many more times?' I asked, looking over my shoulder at her daughter.

'Aye, twice since you were here last. But maybe it's slowing.' I pulled up the elastic on her pyjama trousers and looked down on the clots of blood that slipped out of her like leeches.

'I'll come back before finishing my shift around breakfast time,' I said. 'Try to get some sleep.'

WHEN I RETURNED to the Stafford house it was just before

eight. The bin lorries were out, and the rain was easing. It took a while for the door to be answered.

'Well, she's still breathing,' was the first thing her daughter said, stepping aside to let me in. 'But only just,' added the granddaughter, sitting back and stroking the tense, swollen skin of her belly. 'She's said nothing since you left.'

Her son was asleep on the sofa, snoring. His slippers were placed neatly beside the Pekingese ashtray. The television was still on, but muted. I pushed open the door to the bedroom for the third time that night. Her breathing was deep and steady, and her face seemed to have even less colour in it, despite the natural light now falling in through the window. 'Did the bleeding stop?' I asked, 'I mean, have you had to change many more pads?'

'Just the one after you left,' her granddaughter said, 'I've not needed to since. Is that a good sign?'

'Sometimes,' I said.

Her pulse was even thinner than before – I could barely feel it. Her breath was deep, sighing and sporadic. Her eyes were half-lidded, and grey crusts of spittle had gathered at the angles of her mouth. The creases of her wrinkles seemed smoother, and the tone of her skin had yellowed from wax to something more like old vellum. I was standing holding her wrist, feeling for her pulse, when she made a long, rattling sigh, then fell silent. I stood still for a few moments, out of respect, before glancing down at my wristwatch to count. One minute passed, then two.

'That's it, isn't it?' her daughter asked.

'Yes,' I said. 'She's gone.'

And she began to sob, but silently, showing only in the shudder of her shoulders and the way she rocked on her chair. Her own daughter put an arm around her shoulders, and pulled her close.

16

Afterbirth: Eat it, Burn it, Bury it under a Tree

One can see what custom can do, and Pindar, in my opinion, was right when he called it 'king of all'.

Herodotus, *The Histories*

AT FIRST GLANCE umbilical cords seem to come from the sea: opalescent and rubbery like jellyfish fronds or stems of kelp. Their contours are torqued in a triple helix of blood vessels; twinned arteries spiralled around a single vein. The purplish blood vessels braid themselves through greyish jelly composed of a substance used in only one other place in the body: the refractive humours of the eye. They look soft and delicate but are tougher than appearances suggest; for nine months they have to tether a baby to life.

The wrinkle-faced, bunch-fisted girl I had just delivered was already squalling, and I dried her with a towel and held her down beneath the level of her mother's hips for a moment. The placenta was still inside her mother's pelvis – in these first moments I wanted to let blood run from it down into the baby's body. I put my fingers again on the cord, feeling the pulse of her tiny heart fluttering within it like a trapped moth. 'Is everything alright?' asked her father. He looked stunned by sleeplessness, and the agonies of labour that he had just witnessed his wife going through.

'Fine,' I said, 'absolutely fine.' As I watched the girl, my fingers on her cord, the pulse in it thinned out and then stopped – a reaction to the coolness of the air and the higher oxygen levels in her blood now that she was breathing for herself. Inside her liver and around her heart other blood vessels were closing off in synchrony. These are 'shunts' that during her time in the womb had diverted blood around the developing lungs and liver. Other vessels for carrying blood to and from the lungs were at the same time opening up – it was thanks to them that her blood was reddening with the flush of oxygen. A hole in her heart, necessary for circulation while she was inside the womb, was closing over. Her umbilical arteries were closing too, narrowing from their origins deep in her pelvis running out towards her umbilicus. It was because of this concert of changes that her bluish, waxy face was pinking up. Only when the pulsing of the cord had stopped did I place plastic clips across it.

The midwife handed me some scissors, scored and blunted from their many passages through the steriliser, and once again I marvelled at how a substance seemingly so fragile can be so hard to cut; I had to hack at it as if at a hawser. For the delivery of the baby the mother had been on all fours, but as I handed her daughter up to her she heaved herself onto her back, pulling the baby onto her breast with an astonished gasp. As mother, father and baby dissolved into a universe of three the midwife and I looked down at the business end. This wasn't over yet.

The 'third stage' of labour is unexpected for many, as if the show should be over with the birth of a child. But a storm of hormones and chemistry was shearing the placenta from its mooring against the womb lining. If contraction occurs too slowly the blood can go on pouring from the raw surface of the womb – a 'post-partum haemorrhage'. I pushed my hand gently, but firmly, onto the mother's slackening abdomen, to feel if the womb was shrinking down. It was.

With a pair of steel tongs I pulled gently on the cord. The baby was already at her mother's breast: as she sucked, hormones hastening the let-down of milk also caused the mother's womb to tighten. As I turned the tongs, the cord blenched against their steel – the arteries and vein within were already ghosts of their once vigorous selves. Then, as I pulled, the cord suddenly widened the way a tree trunk does just before its roots spiral into the earth. The 'afterbirth', a violet clot of blood, slithered from the mother's body onto the bed.

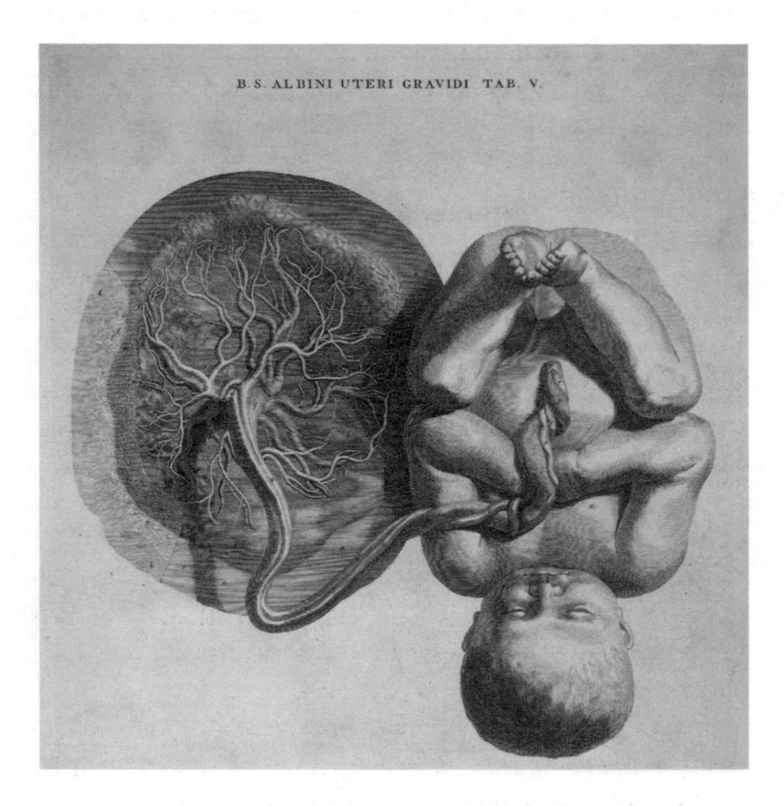

It was heavy – over half a kilogramme – almost round and about an inch thick. Since early in the pregnancy it'd had to carry oxygen, sugar and nutrients towards the developing foetus, as well as carrying carbon dioxide, urea and other by-products back towards the mother. The pressure pulse of this remarkable exchange had been driven by the baby's developing heart. The blood of mother and baby don't mix, but the capillaries belonging to each are brought together so closely that it's as if a million tiny hands locked fingers across the placental divide. Da Vinci noticed this distinction over five hundred years ago, when many of his contemporaries still believed that babies grew by consuming their mother's menstrual blood. Leonardo's placental drawings betray a familiarity with the afterbirth of the sheep; it's thought he only saw one cadaver of a woman who died in pregnancy. He wasn't alone: European men through the centuries seem to have had more familiarity with sheep placentas than those of their own children. Even the scientists' word for the placental membrane, *amnion*, is taken from the Latin for 'lamb'.

Most of the elements of our anatomy are robust enough to see us through four or five decades at least before they start to fail, but an organ that need last only eight or nine months shows just how fragile human tissue can be. I've seen placentas turn brittle and grey, either because of the toxins they've been exposed to or from the relentless deep-fry nature of the Scottish diet. The worst are the placentas of heavy smokers, clotted with knots, yellow and hard as ambergris.

This placenta was clean though, and I spread it out on a steel tray. The gossamer remnants of the amniotic sac were fused into the placenta itself, and I couldn't find any rips. 'Membranes intact,' I said to the midwife, before taking it by the stump and hoisting it awkwardly into a plastic bucket. I clipped on an orange lid as if sealing a pot of paint, then carried it through to the waste room on the ward. From being the centre of this baby's world, essential for life and growth, it was now part of the anonymous stack of placentas and umbilical cords that had been delivered that day, and which later tomorrow would be burned in the furnace beneath the hospital's smokestack. What was nourishing the baby only that morning would tomorrow be smoke floating over the city.

THE GREEK '*OMPHALOS*' comes from the same root as the Latin '*umbilicus*': both carry the sense of being at the centre either of the body or the world. For the Greeks the Omphalos, a stone at the Delphic Oracle, was considered the geographic centre of the earth. Around the time that people were making pilgrimages to Delphi, the Greek traveller and historian Herodotus wrote about the way different customs were prevalent across different parts of the ancient world:

> One might recall, for example, an anecdote of Darius. When he was king of Persia, he summoned the Greeks who happened to be present at his court, and asked them what they would take to eat the dead bodies of their fathers. They replied that they would not do it for any money in the world. Later, in the presence of the Greeks, and through an interpreter, so that they could understand what was said, he asked some Indians of the tribe called Callatiae, who do in fact eat their parents' dead bodies, what they would take to burn them. They uttered a cry of horror and forbade him to mention such a dreadful thing.

For Herodotus, custom was everything, and for the past few decades in the West our custom has been to burn placentas with the soiled dressings, diseased organs and contaminated needles of the hospital incinerator.

Just as Darius' Greeks were horrified at the prospect of eating their fathers, and the Callatiae of India were horrified at the dishonour of *not* eating them, the practice of eating placentas arouses fierce emotions, both for and against. Placentas are a rich source of progesterone, the hormone that maintains pregnancy, and a crash in body progesterone has been proposed as a trigger for 'baby blues' – the disturbance in mood after childbirth that often cedes to post-natal depression. Eating the afterbirth is a common habit among carnivores, as well as omnivores like chimpanzees – our closest relatives. It could be that the practice is not just about nutrition, but about letting an exhausted mother come down gently from her progesterone-high.

There is only one reference in the Old Testament to the afterbirth, and it's about breaking taboos: in Deuteronomy 28, verse 57, a woman is given permission to eat the ordinarily prohibited placenta because her city is under siege. But in other cultures around the Mediterranean rim a new mother was traditionally encouraged to eat afterbirth in order to help her milk come through, and to reduce after pains as her womb contracted back down to its normal size.

From Morocco to Moravia to Java, women have eaten the placentas of their own children, or those of other women, in order to improve their fertility, while in Hungary the ashes of a burnt placenta were fed in secret to men in order to *reduce* their fertility (this isn't as daft as it sounds: female sex hormones can sometimes aid female fertility, while at the same time inhibiting sperm production if taken by men). During the Tang dynasty in China, around the seventh century CE, the placenta of a live-born girl was advocated in a spell to transform oneself into a young girl.*

The eggs of the earliest vertebrates evolved to grow bathed in seawater, and by developing a womb full of amniotic fluid, we mammals have evolved a way of carrying a sea inside us. That the membranes in the womb have a close connection with the sea seems to have been recognised since earliest times: those membranes, the caul, have often been considered protective against drowning. In the cultures of the British Isles, a baby who emerged still wrapped in its caul was destined to be a strong swimmer, and would be possessed of good fortune. Charles Dickens's David Copperfield starts out his autobiography with a discomfiting

* The spell-book was called *Collection of 10,000 Feats of Magic*.

discussion of how his own caul was put up for sale to the highest bidder for just this reason:

> I was born with a caul, which was advertised for sale, in the newspapers, at the low price of fifteen guineas. Whether sea-going people were short of money about that time, or were short of faith and preferred cork jackets, I don't know; all I know is, that there was but one solitary bidding.

As far apart as Japan and Iceland the traditional method of disposing of a placenta was not to inter it beneath a tree, but to bury it under your house. In Japan a priest would choose the location of the burial place, whereas in Iceland it would be buried in such a position that the mother's first steps in the morning, as she rose from bed, would stride across it. Another old Chinese text advised burying the placenta and cord deep in the ground 'with earth piled up over it carefully, in order that the child may be ensured a long life. If it is devoured by a swine or dog, the child loses its intellect; if insects or ants eat it, the child becomes scrofulous; if crows or magpies swallow it, the child will have an abrupt or violent death; if it is cast into the fire, the child incurs running sores.'

The Russians traditionally saw the placenta and umbilical cord as sacred; Orthodox Christians dedicated it in particular to the Virgin Mary, the governess of fertility. Following delivery, the afterbirth would be laid out for a time on the local church altar, where it was believed to influence the fertility of other women in the community, before being buried.

Some Indonesian peoples held that because the placenta and its membranes seem to have come from the sea, it must be returned to it: after being placed in a pot it would be thrown into the river to float back to the ocean. This was done to prevent the placenta falling into malign hands (the idea that the placenta is part of the child, and is in some way identical with it, is a persistent one). Other South East Asian peoples prepared a funerary bier for the placenta, surrounding it with oil lamps, fruit and flowers, before floating it downriver.

For some cultures it has not been the affinity of afterbirth with the sea that has been celebrated, but its resemblance to a tree: the way that the spiral trunk of the cord seems rooted into the earth of the womb. I've been told that during the delivery of a baby – the second stage of labour – the pain women experience is that of relentless waves of pressure combined with the knife-and-fire stretching of the perineum. Passing the afterbirth is quite different; a deep sense of an uprooting, of something long buried being tugged free. In *The Golden Bough*, James Frazer's magisterial work of cultural anthropology, several cultures are described as burying the placenta beneath a sacred or significant tree, which then retains its connection to the child throughout the life of both. The tree is renamed for the child, and becomes the centre of its world just as the Omphalos in Delphi was the centre of their world.

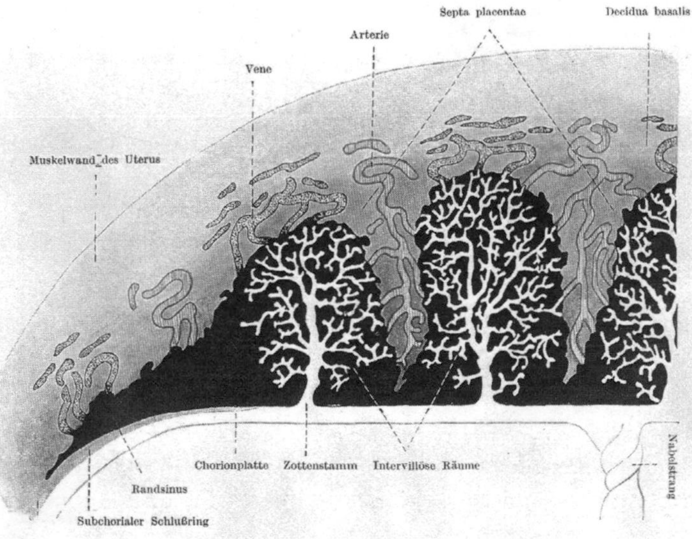

FOR MOST OF us, the landscape of our childhood retains a special power; it's a common experience for its influence to be formative, and extend into adult life. In the West we don't usually sanctify that landscape by burying placentas in it, or by dedicating the afterbirth to a local fertility goddess, but it can carry a sense of the sacred all the same. In the late 1970s Seamus Heaney read an essay on BBC Radio, 'Mossbawm', in which he described the yard of the farmhouse where he grew up in those terms. He called the opening sequence 'Omphalos', and described in it how the water pump behind his back door was the centre of his world as a child. American military were conducting manoeuvres across County Derry, bombers flew low into a nearby airbase, but the rhythms of home remained undisturbed by great historical events. The drone of the bombers was distant; closer at hand was the sound of water falling into buckets, repeating *omphalos, omphalos, omphalos,* as women pumped water for five households from a single pump. The omphalos was the still point at the centre of his life – still but flowing with the water of life, and sustaining the lives of all who lived around it. Heaney was anchored by the water pump, just as the umbilical cord anchors a baby through its nine months in the womb.

In his radio broadcast Heaney wasn't content to meditate just on the pump – his circuit of the sacred landscape of his childhood widened, taking in a field of peas ('a green web, a caul of veined light'), and then a hedge, the fork of a beech, a hay byre, and the hollow trunk of an old willow. The willow was his favourite: he'd lay his forehead against the bark and feel the crown of the tree swaying above him, both embraced by the wood and shouldering it the way Atlas shouldered the world. Then, in a sudden mid-sentence switch of mythology, he recalled the antlered branches rising from his head as if he were Cerunnos, one of the gods of the Celtic pantheon. The landscape was sacred, *omphalos* and caul, and it was irrelevant whether

Christian, Greek or Celtic traditions were used to express that sanctity.

WE CAN EAT IT, burn it, float it on a raft or inter it under a tree. We can call a priest to help us bury it under the house. We can sell it to the highest bidder, drop it into a hole at high tide or secrete it away from bad spirits. In wealthy, modern healthcare systems, a new possibility has arisen: to have it cryogenically preserved.

Buried in the jelly of the umbilical cord are cells that are genetically identical to the baby, but which have not differentiated into any particular tissue type. These 'undifferentiated' cells are a type of 'stem cell' because, just as it's possible to regrow a tree from a single cutting, they are stems from which other body parts can theoretically grow. The cells within the cord blood have the potential to develop into tissues such as bone marrow, while the cells within the jelly of the cord are related to the structural components of the body: bone, muscle, cartilage and fat.

The leaflets advertising umbilical cryogenics have two types of picture on them: cute and smiling children at play, or radiation-suited scientists engaged in some challenging laboratory task. There are no images representing multiple sclerosis, Parkinson's disease or leukaemia, despite the companies' claim that storing stem cells might be an insurance policy against these illnesses in later life. You can donate stem cells to a public bank, for use by anyone, or you can pay a private company to store your baby's cord and stem cells for the sole use of your family.

Some cultures maintain that a baby's visceral connection to its umbilical cord is an association that lasts a lifetime, and for that reason the cord must always be handled with respect. These cryogenics companies agree: if you want a private cord-bank to store your baby's umbilical cord you can arrange for a lab scientist to be on standby for the birth of your child in order to extract the stem cells within the critical time period in which they're still viable. Your baby's lifetime's association with the cord can be maintained through regular payments from a credit card. The National Health Service in the UK now has a cord-blood storage service, preserving stem cells for research, and investigating their use in bone marrow transplants for whoever might need them. Within a decade we've gone from throwing afterbirth out with the trash, to reinvesting it with a depth of significance that had almost been forgotten.

There's some debate as to whether the private banks can ever supply enough stem cells to treat an adult, and so it remains controversial whether the high costs of preserving a child's cord for its own use are justified. While the East African might feel tied to his umbilical tree, rooting him to a particular patch of the earth, you're unlikely to draw strength and a sense of belonging from regular visits to a cryogenics lab. The laboratories themselves share specimens, and your cord may end up being stored in another country altogether, inaccessible to you or your child. But at least it will be out of reach of ants, swine, dogs and magpies.

LOWER LIMB

17

Hip: Jacob & the Angel

His hips were titanium-vanadium,
where the angel touched.

Iain Bamforth, 'Unsystematic Anatomy'

THE HIP IS A STRONG JOINT: a bossed knuckle of bone clasped deep into a hollow of the pelvic skeleton. It's buried beneath layers of the thickest and most powerful muscles in the body. There are four main groups of these, and all of them are active when walking: two groups have their greatest actions on the hip and two groups have their greater actions on the knee. The process of taking a step involves countless adjustments, each muscle continuously testing itself against the strength of all the others. Each movement must take into account uneven terrain, movements of the trunk, and the balance and kinetics of the other leg.

There's a novel by German-Italian writer Italo Svevo in which the protagonist, a hypochondriac businessman called Zeno (after the Greek philosopher of paradox), meets an old school friend whom he hasn't seen for years. The friend is afflicted with debilitating arthritis, and Zeno is surprised to see him walking with a crutch. 'He had studied the anatomy of the leg and the foot,' says Zeno, 'and laughingly told me that when you walk quickly, the time taken to make each step is less than half a second, and within that half-second no less than fifty-four muscles are in motion.' Zeno is horrified by this 'monstrous machinery' in his leg, and he at once turns his awareness inwards, hoping to sense each one of the fifty-four moving parts. The deeper consciousness he obtains doesn't help him to a greater understanding of his body; instead, he becomes baffled by his own complexity. 'Walking became a difficult labour, and also painful,' Svevo wrote. 'Even today, while I write, if someone watches me in motion, the fifty-four movements become too much, and I'm at risk of falling over.' The hip and its movements become so fundamental to Zeno's sense of himself that all he has to do is think of them, and he's immobilised.

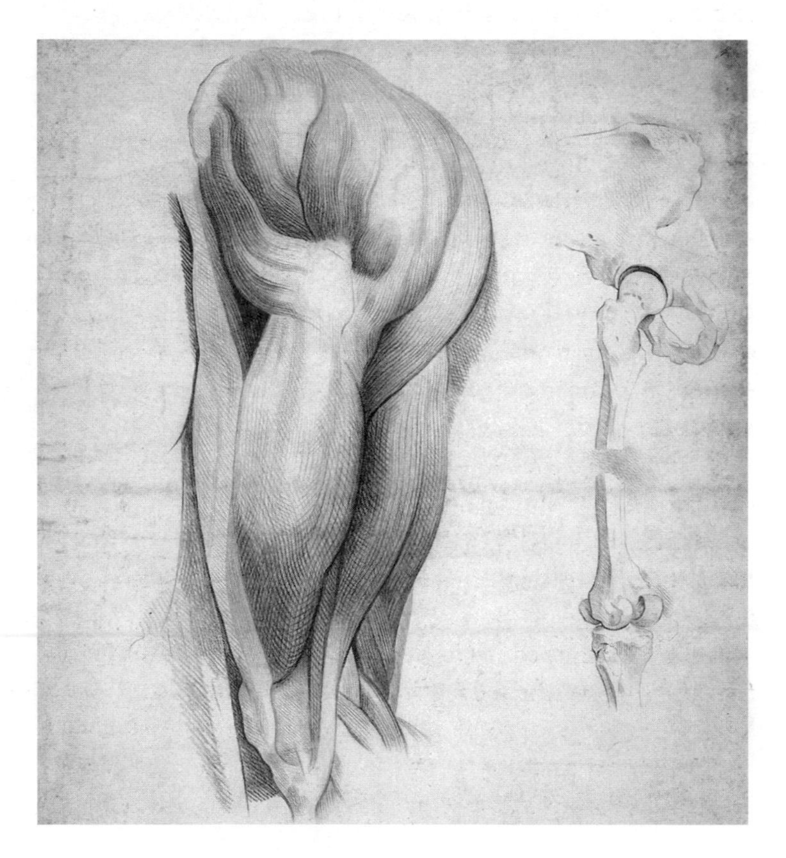

Hips cause all sorts of problems, and seemingly minor issues in childhood can cause a permanent limp if they're not addressed. As foetuses in the womb we fit best if we tuck our legs up cross-legged; if hips aren't flexed in this way they grow with roughened and shallow sockets ('developmental dysplasia'). Once the baby begins to stand up, the development of walking will be painful and slow. I check every newborn baby I see for this problem: grasping the baby's legs I fit each knee snugly into the palm of my hands, and place my fingertips over his or her hips. Pushing down on the knees, and splaying the thighs out and in, occasionally reveals a subtle and ominous click. The cure is straightforward if demanding both for the baby and the parents: both legs must be spread widely and immobilised in a plaster cast for the first few months of life.

After a year or two, another problem may occur in the growing hip: toddlers suffering from viral infections can develop an isolated build-up of fluid within the joint. They start to limp and fall over; these 'irritable hips' settle down without treatment over the course of a few weeks. By the time children are five or six yet another problem may develop: a disruption of blood flow causes a softening and distortion of the bossed head of the femur. This is 'osteochondritis', four times commoner in boys than in girls, and surgery is often required to restore the shape of the bone within the hip.

Once the danger years of osteochondritis are past, and children reach adolescence, they may develop a fourth hip problem: between the ball of the hip joint and the femur itself there's a growth plate within the bone which allows for lengthening of the thigh. This can sometimes detach and slip – a 'slipped upper femoral epiphysis' – and if not fixed by surgery the teenager can be left with a permanent limp.

One of my anatomy tutors used to say that the best evidence for evolution over creationism is how many failings we have – the human body could be far better designed. Much of the misery our hips inflict is a result of their meagre blood supply. There are plenty of places in the body with a blood supply greater than that which is actually needed – you can block off an artery to the stomach, hand, scalp or knee with little consequence. But the hip is far more vulnerable: in common with the eye, brain and heart it has a blood supply which is easily obstructed. Blockage of blood to the brain results in a stroke, to the eye results in blindness, to the heart results in a heart attack. Loss of blood to the hip can be just as catastrophic – even fatal.

If someone over the age of seventy-five falls heavily onto the hip, they've about a one in ten chance of breaking the bone. A crack in the hip often cuts off the blood supply to the ball of the joint, and the bone within that ball dies off. These fractures can't be repaired; the only solution is to cut out the joint and replace it with an artificial one. Debilitated elderly men and women, already so weak they've begun to fall over, frequently struggle to recuperate from such a major operation. Around 40 per cent of them will end up in a nursing home because of the fall, and 20 per cent will never walk again. Between 5 and 8 per cent will die within three months of the fall.

THE HIP CAN REPRESENT the life that as human beings we carry within us. Tibetan Buddhists make trumpets from the bone in order to remind themselves of death, and in the book of Genesis the joint is taken as one of the principal sources of human life. Jacob, grandson of Abraham, fools his brother Esau into forfeiting his inheritance. The two are twins and this isn't their first fight: earlier in Genesis we're told that Jacob was born grasping at his brother's heel (his name Yaakov is related to the Hebrew *akev*, meaning 'heel').

At the outset of the story, Jacob has prepared hundreds of animals as an appeasement gift for Esau. Before he can offer them to his brother he is set upon by an angelic figure who wrestles him to the ground. The *Zohar*, a mystical, Kabbalistic commentary on the first books of the Bible, describes the assailant as representative of humanity's darker side, and Jacob's fight with him as an allegory of the struggle to live a morally upright life. The two fight 'until the break of day', with Jacob trying to extract a blessing from the figure. When the angel realises that he cannot match Jacob fairly, he forcibly ends the fight by dislocating Jacob's hip, leaving him with a permanent limp as a reminder of the night that he took on an angel and almost won. The chapter closes with the newly named Israel's proclamation that he has seen 'the face of God', and explaining that the 'sinews' over the animal hip are henceforth a forbidden food for Jews: 'because he touched the socket of Jacob's hip on the sinew'.

Rabbis and Hebrew scholars can't agree on the exact significance of the story. One perspective is that the hip and thigh were, for the ancient Semitic culture of Abraham and Jacob, storehouses of sexual and creative energy. The word in the text, *yarech*, could refer to the inner curve of the thigh where it folds onto the scrotum in men, and the vulva in women – a Hebrew scholar told me that it is probably better translated as 'groin'. The same word is used in the book of Jonah to describe the inner hollow of a boat, and in Genesis chapter 24 Abraham asks his servant to swear an oath by touching him in the hollow of the thigh – a reference to the ancient custom of swearing by the testes (hence, 'testify'). From this perspective, by touching Jacob's groin and hip the angel imparted the strength and authority to father a whole nation.

There's a rival theological position that claims Jacob's subsequent limp to be the most important factor in the parable: his injury is a reminder that the Jews should not try to stand alone. Jacob tried to fight an angel and, because he was human, he failed. His limp branded him as vulnerable and mortal, as we all are. From this perspective, the strength and progress of the Jewish people depends on an acknowledgement that God decides whether we fail or prevail, live or die.

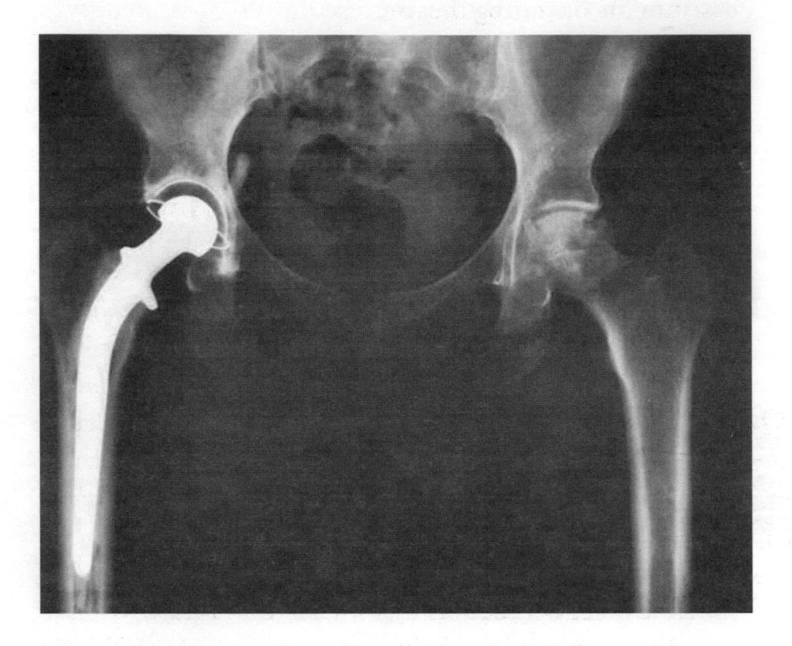

THE FIRST ON-CALL I ever worked in a hospital was a fifty-four-hour shift covering orthopaedics. Before that shift I'd never gone even twenty-four hours without sleep, and my recollection of those hours is hazy and hallucinatory; a delirium of sleep deprivation and panic. At my graduation from medical school a couple of weeks earlier I'd been awarded a gold medal, and handed a certificate awarding me 'MBChB with Honours'. Gold medal or not, it was immediately apparent just how much I still had to learn.

People quickly became their diagnoses. I admitted broken ankles, snapped wrists, dislocated shoulders and crushed spines – each individual had to have paperwork filled in, their X-rays and blood tests arranged, and if they needed an operation then I was to explain the risks of surgery and have them sign a disclaimer that they accepted those risks. At the same time, there were two wards full of patients who needed checked and attended to, hundreds of drugs and intravenous fluids to prescribe, and my boss to assist in the operating theatre.

One of the first patients I ever admitted was Rachel Labanovska, a 'fractured neck of femur' according to my new, technical language, but in human terms an eighty-four-year-old lady who ordinarily lived comfortably and alone, managing all her own affairs, though she required the help of a metal walking frame. Some years before she'd fallen and fractured her left hip: it had been replaced by a metal alloy one which had succeeded in helping her maintain some liberty and independence. A few days before I met her she developed a chest infection – her daughter had noticed a cough – and her family doctor prescribed some antibiotics. The antibiotics didn't work well enough and she became feverish and delirious, falling over her metal frame and breaking her other hip. She lay on the kitchen floor for eighteen hours before her daughter found her; by the time I met her she was hypothermic and close to death.

She lay on a gurney hallucinating, her limbs stick-thin, waving her fingers in the air as if each was a magic wand. Her right leg was shorter than it should have been, and her knee was facing out to one side: 'shortened and externally rotated', as the textbooks put it. When I attempted to take blood from her arms the dreaminess vanished: she dug her fingernails into my skin and shrieked as if being disembowelled. I had to hold her down to take blood and, because her temperature was still dangerously low, sedate her so that she'd stay put beneath the hot air blanket we'd set up to warm her.

Mrs Labanovska was trapped in a terrible paradox: without surgery to replace her hip she'd be killed by her pneumonia, but because of the infection in her lungs she was too weak to survive surgery. I took her daughter to one side to explain. Hope, fear and anxiety moved across her face like cloud shadows. 'So what now?' she asked me. 'My mother is a feisty lady – she's travelled all over the world. She couldn't cope with being dependent on others, living in a nursing home.'

'We'll take her upstairs and give her strong antibiotics,' I said. 'You say she's a fighter – she may recover enough for the operation.'

She was taken to a side room on the orthopaedic ward where I set up intravenous antibiotics, a mask giving high-flow oxygen (which, in her confusion, she kept pulling off), and arranged for a physiotherapist to help her to cough mucus from her lungs in order to improve her breathing.

I've seen death come as meekly as an expiring candle, or terrible and all-consuming – a black star. Mrs Labanovska was tiny and wizened, but her life had been daring and expansive, and her death was equal to its drama. For the first few hours she was quiescent, only muttering if she was disturbed by me, the nurses or the physiotherapists. Then the delirium caused by her infection took greater hold: confusion laden with fury began to thicken in her mind. She tried again and again to leave her bed, but howled with agony whenever she tried to move her broken hip. She was unable to stand. At some point in the middle of the first night her daughter went home to rest and was replaced by a son, who sat by her bed while she writhed and moaned. I gave morphine for her pain, but too much would hasten her death, and there was still the chance that she might survive and be able to undergo surgery.

On rounds the following morning, twenty-four hours into the shift, the surgeon in charge explained to her son that the next few hours were critical: if her breathing did not improve she would be unlikely to survive another night. Mrs Labanovska's pulse by that time was what they call 'galloping': a stampede towards oblivion. She still shrieked if she was moved, but had given up trying to escape her bed. Through the day I tried to visit her room, to talk to her expanding number of relatives, but it was midnight on the second day before I had the chance. She was peaceful, then: though her breath came fitfully she was less tormented by her struggle with both the pneumonia and with her broken hip.

During lunch the following day with my colleagues I was blurry-eyed with exhaustion when my bleeper squealed once more. 'It's Mrs Labanovska,' said the nurse on the end of the phone. 'She's dead. Do you want to certify her, or shall I get someone else to do it?'

'What was that?' asked the registrar as I put down the phone.

'Mrs Labanovska's dead. I've got to go down and certify her.'

'Don't rush,' he said through a mouthful of food. 'Let the poor woman get cold first.'

WHEN I ARRIVED at the ward her family were gathered outside the room. The nurses had laid her out neatly, and made the deathbed up with clean sheets. As I listened for a heartbeat that didn't come, and shone a light into eyes that didn't see, I glanced down at the shortened, rotated leg that had killed her.

If someone is to be cremated rather than buried there are two forms to be filled in by the attending doctor: the death certificate, and the cremation form. The cremation form certifies that there were no suspicious features surrounding the death, and so incinerating the body won't destroy evidence. The other function it serves is to reassure the undertakers that there are no pacemakers or radioactive implants in the body. Pacemakers can explode when subjected to the heat of a cremator, and radioactive implants, which are used in the control of some cancers, are dangerous to others if left among the ashes.

'She's for cremation,' the nurse in charge said, handing me the form. I stood in the middle of the ward, with Mrs

Labanovska's daughter and son standing beside me, answering the bleak, bureaucratic questions while porters hurried by with other patients and phones rang unanswered on the desk. 'Have you, so far as you are aware, any pecuniary interest in the death of the deceased?' NO. 'Have you any reason to suspect that the death of the deceased was due to: a) Violence, b) Poison, c) Privation or neglect?' NO, NO, NO. 'Have you any reason whatever to suppose a further examination of the body to be desirable?' NO. Then I had to sign the certificate 'on soul and conscience'; the final words picked out in red, as if in letters of fire.

'Gosh!' said her daughter, suddenly. 'What about the other hip?'

'Sorry?'

'Her left hip, the one they replaced. It's made of metal. What'll happen when it's cremated?'

'Don't worry about it,' I said, 'the crematorium will sort it out for you.'

CREMATORIUMS ASK RELATIVES if they'd like the metal body parts of their loved ones returned to them, or sent on for recycling. Prosthetic hips, knees and shoulders contain some of the most high-performance alloys yet devised: combinations of titanium, chromium and cobalt that, after gifting mobility and independence to the elderly in their later years, are collected by the crematorium, melted down, and turned into precision parts for the engineering of satellites, wind turbines and aeroplane engines.

THERE'S AN ENDURING FASCINATION with Jacob's struggle because he seems to be wrestling not just with an angel, but also with the frailty and resilience that as human beings we all embody. Some commentators have gone so far as to see in it all the hallmarks of a classic folk tale, in which an individual embarks on a perilous journey, takes on forces that seek to destroy him, is branded by that struggle, but ultimately triumphs. It's a pattern that mirrors the convalescence stories going on in orthopaedic and rehabilitation wards all over the world – journeys like the one that Rachel Labanovska made when she fractured her left hip and had it successfully replaced, an experience by which she was marked but from which she recovered.

Some of the most enduring myths have several layers of possible interpretation and features that resonate across cultures. Some conclude naturally with the victory of the hero, but though they conform to patterns, not all of them have happy endings. In Genesis, Jacob makes it to a new homeland in Canaan, but is swept on by the narrative to Egypt. He dies there many years later, an old, troubled man. Genesis chapter 49 sees him distribute blessings – some barbed, some bountiful – between his twelve sons. Then, 'when Jacob had made an end of commanding his sons, he gathered up his feet into the bed, and yielded up the ghost' – he wasn't transfigured, or transported to heaven. Rachel Labanovska had a more fitting, mythic end: some part of her lives on, and is even now whirling through the sky as a turbine, or orbiting high over the planet she once explored.

18

Feet & Toes: Footsteps in the Basement

That's one small step for a man, one giant leap for mankind.

Neil Armstrong

OCTOBER IN GRANADA: the old Arab quarter of the Albaysín turns its face south, towards the heat of Africa. The narrow streets and architecture still echo the glory of Moorish Spain. I've come to stay with a friend who lives there in a traditional house – a *carmen granadino*. The walls follow the contours of the hillside: we enter from a door at street level onto an upper floor at the head of the house, then descend by wooden staircases to a living area at its foot. The living room opens onto a south-facing garden.

At the end of the garden there is a shrine – nothing else to call it – built over the grave of a mummified little toe, buried in a tiny coffin. The owner of the house, Chemi, is also the owner of the toe. He lost it in a road traffic accident in 1994, and with the insurance compensation he was able to put down a deposit to buy the old house. He has officially renamed the house Carmen del Meñique – the House of the Little Toe.

Since losing his toe, every year in October Chemi holds a *romería* – a traditional mourning ceremony. The toe is unearthed and paraded through the streets of the city on a decorated platform more used to carrying images of Christ or the Virgin – proclaimed as an uncorrupted relic. There can be as many as two hundred devotees taking part, and as they process around the Albaysín they sing laments, making their way to a consecrated fountain, where they anoint the stump of Chemi's foot then throw a riotous party. After a circuit of the Granadan streets the relic-toe is reinterred for the following year.

The foot is often overlooked by anatomists, relegated to the last pages of the textbooks and the last days of revision for students. But it's said that the anatomical structure of the foot tells us something essential about humanity, about how as apes our ancestors came out of the forests and walked into our modern selves. There was something about the procession of the little toe that struck me too as quintessentially human: the ability to poke fun at a solemn ceremony, and transform pain and loss into a glorious celebration.

IN 1978 THE PALAEOANTHROPOLOGIST Mary Leakey uncovered three sets of ancient footprints on the Laetoli plain of Tanzania. The prints extended for almost ninety feet and seemed those of a man, woman and child, walking together across some moist volcanic ash which had later solidified to rock. Part way across the span of footprints one figure paused for a moment as if in indecision, turned to the left, then carried on. More ash subsequently fell on the footprints and preserved them. It was raining as they walked: the ash had also preserved the imprint of raindrops.

The prints were made more than three and a half million years ago. They were not humans as we know them today but *Australopithecus afarensis*, one of the hominid roots of humanity's family tree. *Australopithecus* had small, gorilla-like brains, did not yet know how to chip stone tools, but unlike gorillas they walked upright as we do. What did the figure stop to watch? It could have been a nearby volcano erupting, the source of the volcanic ash drifting over the plain. Perhaps the group was a family hurrying away from the eruption, from a foreboding, darkening sky. One of the sets of footprints had pressed deeper into the ash with the left foot, as if carrying a baby, a burden, or even struggling with a limp.

Experts in functional anatomy are able to gauge the weight, walking speed and species from these subtle footsteps in the ash: to the non-specialist, the imprints are indistinguishable from our own human ones. Computer simulations based on fossil remains have estimated the group's speed, gait and stride length. Like us, *Australopithecus afarensis* had big toes in line with the other toes, had arches to their feet, and they walked by striking the ground with the heel bone (calcaneum) first then pushing off with the toes. Before the Laetoli footprints were found, it was thought that an increase in brain size occurred before hominids began to walk upright, but Laetoli proved otherwise: that it was only by learning to walk upright that we freed our brains, and our hands, to manipulate abstract concepts and the raw materials of the world.

MEDICAL STUDENTS may learn the anatomy of the foot last and pay it little attention, but the foot is a marvel of engineering – when we run, around half of all the energy used in each step is stored in the elasticity of our Achilles tendons and sprung into the arches of the feet. The shape of our footprints is a reflection of three arches that bear our weight: two along each foot's length, and one across its width. Parents worry about 'fallen arches' in their children not just because they look odd, but because they can lead to pain and disfigurement. Like the arrangement of spans on a bridge, the arches of the foot are necessary for strength: without them, the foot cannot adequately bear the weight of our bodies.

The arches of the foot are sustained in four ways. There are bones at the apex of each of the three arches that are shaped like keystones, with their wedged surfaces pointing groundwards. There are ligaments joining each bone along its under-surface, like the staples running between each stone on the underside of a bridge. Tendons and tough, longer ligaments run from one side of the arch to another, like tie beams strapping the two ends of a span. Other tendons anchored in the leg hang the arches like the cables of a suspension bridge.

The neglect of the anatomy of the foot is undeserved. If we believe the evidence of the Laetoli footprints it's thanks to the arches of our feet that we stepped into our humanity.

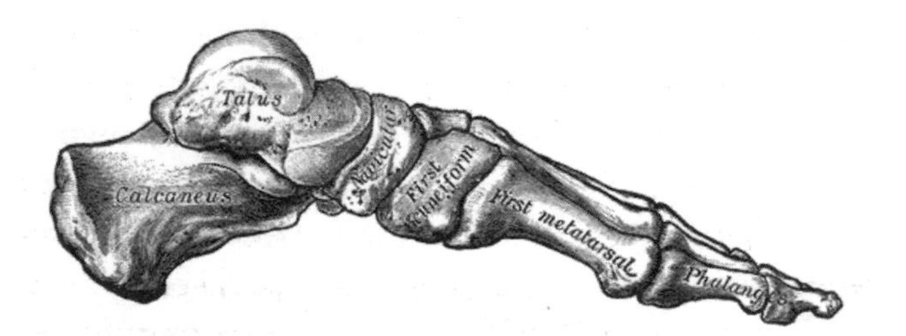

Walking for too long, or burdened by too much weight, leads to stress fractures in the metatarsal bones just like cracks in the stone of an overstressed bridge (these are called 'march fractures' because they were first recognised in soldiers on the march). Ligaments that strap the arch in place can become irritated and inflamed; this 'plantar fasciitis' can be tormenting, and difficult to settle. Gout often attacks the joint at the ball of the foot, and Morton's neuromas – painful swellings that develop on nerves – often develop in the web spaces between the toes. Children with flat feet need assessment for insteps or even special shoes, so that their bones can grow into a supportive arch. Even if scant attention is paid to feet at medical school, qualified doctors have little choice but to spend time thinking about their anatomy, and how to heal them when they go wrong.

ONE OF MY ANATOMY LECTURERS was Gordon Findlater: a straight-talking Aberdonian with quick hands and a silver beard. Before he became an anatomist, Gordon had worked as a telephone engineer. Perhaps he was a natural teacher, or perhaps his work fixing telephones had given him a talent for communication. He asked us, 'Which is more specialised in terms of its function, and specific to human beings: the hand or the foot?'

'The hand!' we shouted. 'Opposable thumbs!'

'Wrong,' he said, explaining that opposable thumbs are an easy modification, only subtly different from the hand of apes. 'It's the foot that is adapted to walking upright,' he said, 'the foot is more specific to us humans.'

I used to work for Gordon preparing dissections of human anatomy as demonstrations for students. Up in the dissection room, high, draughty ceilings were supported by elaborate cast-iron beams, and for most of the year a cold, bleaching light fell in through its north-facing windows. Perched on a high stool I worked on trays of body parts, or sometimes a full cadaver. It was restful, meditative work, equally occupying hands and mind. It was also revelatory: I found myself struck by a sense of wonder at the intricacy of our physical selves. There was satisfaction when a difficult anatomical arrangement was revealed: the brachial plexus, say, or the course of a pelvic artery. On dissecting out the pulley systems of tendons and nerves that control the fingers, I'd marvel that those same mechanisms were enabling my fingers to perform the work.

The dissections I prepared were often of individual body parts: hands, feet, legs, arms, faces or chests. Each part had a plastic tag identifying it – legislation ensures that a record is kept of all body parts dissected, and each was tagged so that the bodies could be reunited, after a fashion, for later cremation. The parts were wrapped in preservative-soaked rags and kept in large bins on wheels – a separate bin for each segment of the body. I'd sometimes go down to the basement and collect the appropriate bin for the day's dissections.

The foot of the building was accessed by an old elevator. In dimensions the elevator was not deep, but it was wide enough to slide a coffin in sideways. Once you stepped inside, you had to pull across a black metal grille, of the same vintage as the ceiling beams, and slam it shut or the lock wouldn't catch. If you were sharing the elevator with one of the cadavers you'd have to hold your breath against the smell, and draw yourself into one corner to make space. Then, with the push of a button, you'd descend into darkness.

The base of the shaft opened into the embalming room: white-tiled walls, terracotta floors, the sharp smell of preserving fluids in the air. There were two embalming tables of mirrored stainless steel; each had two half-panels, which met in a 'V'-shaped gutter. Alan the embalmer was kind-hearted and hard-drinking, with pickled skin and bottle-bottom glasses. He had been an undertaker once, but was glad to be rid of all the velvet curtains, hearses and flower arrangements that become necessary when you work not just with the dead, but with their relatives too. He'd been in the Army Reserves during the first Gulf War, and told me that it was the faces of Iraqi dead that haunted him, not the ones he embalmed. He kept a bottle of whisky on the high shelf of the mortuary office, and drank in a pub called 'The Gravediggers'.

The cadavers were often of people who'd benefitted from the local hospital and wanted to find a way to give something back. Soon after they arrived he'd lay them out on the embalming table and cut down to the femoral artery in the groin, or sometimes the carotid artery in the neck. After inserting a metal cannula he'd attach a rubber hose and strap the cannula in place with twine. A barrel of preservative solution was suspended from the ceiling – after connecting the barrel to the rubber hose, gravity would do the work of pumping preservatives through the blood vessels. As the fluid seeped into the body, blood would leak from the ears nose and mouth, and drain away into the steel gutters.

Just off the chiller room for the cadavers, where the bins for the body parts were kept, there was a descending ramp that ended at a thick, heavy door. One day, when I'd been working on the anatomy of the foot, I asked Gordon what was through that door. 'Do you want to see for yourself?' he asked, pulling out his keys. 'It's in there that we keep all the stuff there's no space for in the museum on the top floor.'

BEYOND IT THERE WAS DARKNESS. Vaulted arches of brick curved like ribs over narrow passageways. The ceilings were low, the air mineral, but at the same time I had the sense of being swallowed by something organic – drawn into the belly of a whale. Gordon found a switch and, after the buzz and crackle of old fluorescent tubes, the space filled with a dismal yellow light.

Catacomb corridors extended out of sight. They seemed to reach beyond the walls of the medical school, branching away within the earth towards the music school, the lecture theatres, and in the direction of the university's grandest auditorium – the McEwan Hall. Racks of human skeletons were hanging along them, facing onto a pile of boxes labelled 'Maori Remains – For Repatriation'.* Their eyeless sockets made me feel as if I was being watched. It was not just an ossuary, it was a menagerie too: a giraffe skeleton lay in crates beside parts of a hippopotamus. In a long polystyrene case I found two ivory narwhal horns, fissured like antique ceramic. Whale vertebrae lay along the edges of the corridor, as if pushed to one side of a plate. Dusty glass bottles lined the shelves, their labels written in copperplate calligraphy. In one corner an articulated skeleton of an orang-utan was gazing towards the exit.

I stopped to open the topmost of another pile of boxes – they contained corpses dissected and lacquered by Alexander Monro Secundus over two hundred years earlier. Monro was an eighteenth-century professor at Edinburgh who revealed much of the anatomy of the brain. Beside them were torsos which his successors had injected with mercury in order to expose the lymphatic channels that are otherwise too fragile and lucent to see. The hearts, lungs and viscera of the bodies were shrivelled and blackened, as if the corpses had been smoked over a fire. They had been sealed in polythene bags to preserve them: mummies dedicated not to eternal life, but to the dream of understanding the human body and its place in creation.

In a crypt-like chamber there were stacks of foetal bones in little cartons, each bone as delicate as a frond of coral. There was a leathery, desiccated face, collected by anthropologists on the Pacific island of New Britain and donated to the museum; it had been peeled off and embedded with clay in an unrecorded rite. In the eye sockets, the New British had inserted little cowries, which gazed out

---

* The University of Edinburgh has been particularly active in trying to repatriate remains that were insensitively 'collected' in former centuries.

glassily at the brick walls. I found the skeleton of an achondroplastic dwarf, stunted with rickets, its femurs and tibias tangled together like gnarled offcuts of oak, and a fetolith, or 'stone baby', retrieved by surgeons from a woman's belly decades after it had died. On a lower shelf there was a glass case and inside it another curled up, mummified baby. 'I don't know where that came from,' said Gordon, 'it could be two hundred years old.'

In an alcove to one side, with a raised floor and lowered ceilings, Gordon pushed through a door hung with an old tin sign: 'Area D', it said, 'under no circumstances move anything from these shelves'. Inside were racks of what the Renaissance medical texts called 'Monsters and Marvels': aberrations of human development, which the nineteenth-century anatomists had salvaged from the middens and fire grates of the city. Mermaid babies with fused legs floated in chambers of briny fluid, alongside a series of Siamese twins ending in a child with a normal body but two heads. Another shelf illustrated various degrees of 'hydrocephalus', or water on the brain, in which skulls ballooned by fluid pressed hard against the confines of the glass. Foetuses dead by miscarriage over a century ago floated in vitrine wombs, their bones stained scarlet, their bodies translucent as jellyfish. It was macabre, but there were sound reasons for filling these shelves at the time: the babies were studied for clues as to the way embryos develop in the womb. Then as now it was hoped that understanding how development went wrong would inform the likelihood of a particular couple having another baby affected.

Another series of shelves seemed like a recapitulation of everything I had ever studied in anatomy class, all hidden here in the lowest and most overlooked reaches of the Anatomy department: sagging, deflated brains; a wax model of the kidneys; an eyeball cut in section. One shelf was occupied with a series illustrating the cochlea and the semicircular canals. A plaster-cast head demonstrated the muscles of facial expression. A heap of placentas and cauls, poorly preserved, disintegrated in wooden cases. On the top shelf, close to the arched vault of the ceiling, a woman's vulva had been staked out in a shallow chamber, exposing the urethra and the thickened tissue around Skene's glands.

Down near the floor, on the rough pine shelves, a bear skull lay beside the skeleton of a bear's foot. The bear is one of the few mammals able to walk upright, and bears, like humans, strike the ground heel-first as they walk. When Leonardo da Vinci was unable to get hold of a human foot, he dissected instead the foot of a bear. The specimen stood next to a series of human feet, stripped in layers from the skin down to the bony arch.

It was time to get back to work. The lights crackled off, the shelves and skeletons returned to darkness. The door clanged behind us as if shutting a bank vault. I trundled a bin of feet towards the elevator, past the chiller room and the embalming tables. Darkness again, then the close, cold feeling of stone walls and stale air. We stepped into the crisp northern light of the dissecting room as if emerging from a tomb and back into life. 'We've got to do something with all that stuff,' Gordon said as we emerged. 'We can't just leave it down there, hidden from everyone but a few specialists.'

In Granada I had felt strongly how as humans we charge our bodies with meaning, whether funny or solemn, and the basement too had felt charged with meaning. On the shelves was the evidence of two or three centuries of restless intellectual energy: mankind's attempts to make sense of the human body, with the aim of healing it when it fails, and easing suffering. There was wonder there too: walking the catacombs reminded me of something Virginia Woolf once wrote about the mind of Sir Thomas Browne: 'A halo of wonder encircles everything that he sees ...; a chamber stuffed from floor to ceiling with ivory, old iron, broken pots, urns, unicorns' horns, and magic glasses full of emerald lights and blue mystery.' Perhaps to some the basement would be an unsettling space, but there were haloes of wonder there in the darkness. I agreed with Gordon: anatomy is too important, too wonderful, to be hidden away or left to specialists alone.

A FEW YEARS AFTER I first visited the crypt beneath the Anatomy department, an intruder broke into the McEwan Hall next to the old medical school and set off the burglar alarms. The alarm had a direct link to the police station, and officers and a dog team were sent to the scene. With police dogs chasing him, the burglar managed to find a hatch into the basement of the hall, and from there kick his way down a tunnel-like passageway that led beneath street level towards the catacombs.

His route was discerned later by following his footprints, as well as the marks he made on the doors that he managed to kick open. He ran in darkness, feeling his way along the stone walls, with the police dogs close behind him. The first door he kicked open led into an old boiler room, and from there he felt his way to another door. After repeated attempts (testified by the number of boot marks photographed later) he kicked it open, and managed to get into the basement beneath the Anatomy department. Feeling his way through racks of hanging skeletons he passed in darkness the shelves of monsters and marvels, the narwhal horns and the giraffe bones, the staked-out genitals and the dissected bear feet. Some items in the basement were disturbed by his panicked, outstretched hands. At the door towards the embalming room he paused, then spent some time trying to force it open. It's just as well he didn't make it through – he would have found himself in the chiller with the cadavers.

Moments before capture, with the dogs that were pursuing him already in the basement, he caught a glimpse of light at the top of an old coal chute. He groped up to it, squeezing his body into a space as narrow as a coffin, then managed to turn himself round and kick out a grille and escape. He must have been fast: once outside he outran the dogs.

THE FOOT PROVIDES some of the earliest identifying evidence about our origins as human beings, and footprints are the signatures of our passage through the world. We occupy space in the world with our feet, expressed through phrases like 'to set foot', 'get off on the right foot', or even 'one foot in the grave'. The first upright footprints that have been found are those three-million-year-old ones at Laetoli, but now there are footprints in the dusts on the moon, that will outlast all of us. Perhaps one day there'll be footprints on Mars.

There was a time when to be a physician and to study anatomy was all about collecting bits of bodies and bottling them; pinning them to boards and stockpiling them in archives. The scientists who make sense of footprints are following a tradition, reaching back through da Vinci and beyond, of attending to the subtleties of anatomy, and applying that knowledge to fundamental questions of humanity. There's still reason to be grateful to those collection-obsessed anatomists; more and more of their work is coming out of darkness and back into the light.

Epilogue

I bequeath myself to the dirt to grow from the grass I love,
If you want me again look for me under your boot-soles.

Walt Whitman, *Song of Myself*

MY MEDICAL OFFICE is a converted tenement flat on a busy Edinburgh street. The consulting room faces east: on summer mornings it's luminous and warm, and in winter it's sepia-toned and cool. A steel sink is set into one corner beneath cupboards stocked with sample bottles, needles and syringes, while in the other corner is a refrigerator for vaccines. There's an old examination couch behind a curtain, and on it, a pillow and a rolled-up sheet. One wall is lined with bookshelves, while others are decorated with da Vinci's anatomical drawings, noticeboards and certificates from medical specialist colleges. There's a chart of the city marked with the boundaries of the practice – a diagrammatic urban anatomy of coloured motorways, rivers and B-roads.

I journey through the body as I listen to my patients' lungs, manipulate their joints, or gaze in through their pupils, aware not just of each individual and his or her anatomy but the bodies of all those I've examined in the past. All of us have landscapes that we consider special: places that are charged with meaning, for which we feel affection or reverence. The body has become that sort of landscape for me; every inch of it is familiar and carries powerful memories.

Imagining the body as a landscape, or as a mirror of the world that sustains us, can be difficult in the centre of a city. In terms of geography my practice area is relatively narrow – it's still possible to visit all of my patients by bicycle – but the cross section of humanity it encompasses is broad. It takes in streets of opulent wealth as well as housing estates of startling poverty, solid professional quarters as well as the student apartments of a university. To be welcomed equally at the crib of a newborn and in a nursing home, at a four-poster deathbed and in a squalid bedsit, is a rare privilege. My profession is like a passport or skeleton key to open doors ordinarily closed; to stand witness to private suffering and, where possible, ease it. Often even that modest goal is unreachable – for the most part it's not about dramatically saving lives, but quietly, methodically, trying to postpone death.

In the centre of the district, not far from the clinic, is a graveyard that is held apart from the city by a high wall. A gravel path winds between mature stands of birch, oak, sycamore and pine; their roots cradle the coffins as they crumble back to earth. My visits are snatched moments between house calls and clinics, and I usually have the place to myself. Occasionally I meet a parents' group out taking a break, like me, from the noise of the city streets. We smile and nod acknowledgement as we pass; toddlers that I've met in the clinic run laughing between the stones; babies I've checked over are lulled gently to sleep in their prams.

The surnames engraved on those stones are familiar: the same ones appear on my computer screen each day. Some of the memorials have an ostentatious gravitas, while others are modest and simple – just a name and two dates. There's something democratic about the way the rich and poor lie side by side. A row along one wall is reserved for the local Jewish population: it's cordoned by steel railings, but the tree roots break through regardless. There are memorials to those who've died far away in the service of a lost empire, from bullets, childbirth or tropical fevers. Some celebrate venerable lives, while others grieve the calamity of an early death. The occupations of the dead, engraved on the memorials, reveal social shifts over the past century or so: cloth merchants, millers, clergymen, bankers. There's an obelisk to a master-apothecary, erected when he would have mixed his own tinctures, and the gravestone of a physician who once served those lying around him.

Sparrowhawks nest in the treetops, and hunt the mice and small birds that live among the graves. Ivy runs over the fallen stones, and between the slumped earth of the plots are thickets of sweet brambles. Summer brings a sort of silence, thick and lush, and sometimes I fancy that beyond it I can hear the soft breath of the leaves. By autumn, those leaves cover the graves in crimson and gold, then in winter the stones stand like sentinels between drifts of snow. But in springtime the branches thicken with fresh leaves, and shoots of new grass push into dappling patches of light.

Life is a pure flame, and we live
by an invisible sun within us.

Sir Thomas Browne, *Hydriotaphia* (1658)

Acknowledgements

My thanks most of all go to my patients, past, present and future – without them I'd be a mariner without an ocean. The requirement to honour their confidentiality means that they must go unthanked individually, but that in no way makes me less grateful.

Hippocrates in his famous Oath emphasised the importance of paying reverence to all those 'who taught me this art', and I've been lucky in the examples set by my teachers. There have been scores of them, but thanks in particular to Gordon Findlater, Fanney Kristmundsdottir, Khazeh Fananapazir, John Nimmo, Theresa de Swiet, Hamish Wallace, Peter Bloomfield, John Dunn, the late Wilf Treasure, Clare Sander, Tim White, Colin Robertson, Janet Skinner, Philip Robertson, Mads Gilbert, Iain Grant, Sarah Cooper, Colin Mumford, Rustam Al-Shahi, Jon Stone, Ian Whittle, Stephen Owens, Mike Ferguson, Sandy Reid, Catharine George, Charlie Siderfin and Andy Trevett.

The foresight, imagination, editorial skill and trust of Andrew Franklin at Profile, and Kirty Topiwala at Wellcome, have been fundamental from the earliest stages. Cecily Gayford at Profile brought her hawk-eyed diligence to bear; and thanks, too, to Susanne Hillen for her meticulous attention, encyclopaedic research powers and forbearance. I've also been fortunate to receive support from both Creative Scotland and the K. Blundell Trust, in order to swap some time in the clinic for time in the library. Jenny Brown provides a glittering example of what every literary agent should be: thorough, approachable, and a whizz at table tennis.

Jack and Jinty Francis, Dawn Macnamara and Flaviana Preston have helped balance the demands of the clinic, the library and the nursery. Will Whiteley read the manuscript early on, gave some magnificent advice, and showed me new perspectives on the brain. For insights and expertise relating to electroconvulsive therapy I'm grateful to Neil McNamara. To John Berger: immeasurable gratitude and a bottle of Talisker for his open-hearted support and his interest in this book from the start. Selçuk Demirel permitted me to reproduce his *Les Etoiles*. Greg Heath and Hector Chawla helped me navigate the labyrinth of ophthalmology. Robert Macfarlane is an unflagging supporter of the ideas behind this book, as well as denier-in-chief of doubts. Thanks also to Bob Silvers at the *New York Review of Books* for setting me off on a journey to the inner ear. Thanks to Peter Dorward for his wit, attentiveness, expert reading skills, gold-plated reflections, and for indulging me with *The Iliad*. I'm grateful to Tim Dee for being a cheerleader of my work, and for his enthusiasm about the noises of the heart. For the section on the wrist, Reto Schneider kept me right and helped me explore the dark and zealous world of Pierre Barbet. Yves Berger permitted me to reproduce a drawing from his exhibition *Caring*, and gave me the keys and the freedom of Quincy. I would have become tangled in the thorny briars of *Grimm's Tales* without the clear-sighted guidance of Marina Warner and the generosity of Lili Sarnyai. David McDowall let me tell some of his story in the chapter on the kidney. Alec Finlay permitted me to quote from his work *Taigh: A Wilding Garden,* relating to the National Memorial for Organ and Tissue Donors. For the section on the hip, Kurtis Peters was my guide to Hebrew scholarship. David Farrier sprinkled the manuscript with editorial diamond dust, and was the inspiration for the story about Jacob – a true scholar and friend. Adam Nicolson was good enough to talk to me about Zeno and *The Iliad*. Thanks to Paddy Anderson and Chemi Marquez for their incomparable humour, hosting me at the Carmen del Meñique in Granada, and giving me an insight into the *romería*.

Robin Robertson graciously gave me permission to write about and reproduce 'The Halving'. Iain Sinclair permitted me to quote from *Landor's Tower.* Kathleen Jamie and Brigid Collins permitted me to write about, reproduce and quote from *Frissure*. Thanks to Iain Bamforth for letting me quote from 'Unsystematic Anatomy'. David McNeish – scholar, physician and theologian – guided me to the Hartman paper on Jacob and the Angel. Thanks to Douglas Cairns for confirming the rule that classicists have a hilarious sense of humour.

Institutions: Thanks to the National Library of Scotland, the Anatomy department of the University of Edinburgh, the Anatomy Museum and the library of the University of Turin, the Wellcome Trust, the medical school of Pavia, the staff of the Royal Edinburgh Hospital, and the Department of Clinical Neurosciences in Edinburgh.

Versions of 'Neurosurgery of the Soul' and 'On Seagull Murmurs, Ebb & Flow' first appeared as Diary pieces in the *London Review of Books*. I'm grateful to Mary-Kay Wilmers for permitting them to be reproduced here, and Paul Myerscough for editing them so attentively. Some of the material in 'Voodoo & Vertigo' was courtesy of my essay 'The Mysterious World of the Deaf', in the *New York Review of Books*.

My colleagues at Dalkeith Road Medical Practice put up with my inconvenient absences with magnanimity – I am eternally grateful to Teresa Quinn, Fiona Wright, Ishbel White, Janis Blair, Geraldine Fraser, Pearl Ferguson, Jenna Rowbottom, and Nicola Gray.

Saying 'thank you' doesn't adequately express how grateful I am to Esa. She is truly one of life's enthusiasts; in many ways this book is for her.

Notes on Sources

2. Seizures, Sanctity & Psychiatry

p. 19 *'a spiritual disease …'* Hugh Crone, *Paracelsus: The Man Who Defied Medicine* (Melbourne: The Albarello Press, 2004), p. 88.

p. 20 *'first recorded instance …'* R. M. Mowbray, 'Historical Aspects of Electroconvulsive Therapy', *Scottish Medical Journal* 4 (1959), 373–8.

p. 20 *'he claimed a …'* Gabor Gazdag, Istvan Bitter, Gabor S. Ungvari and Brigitta Baran, 'Convulsive therapy turns 75', *BJP* 194 (2009), 387–8.

p. 21 *'there was a feeling …'* See Katherine Angel, 'Defining Psychiatry – Aubrey Lewis's 1938 Report and the Rockefeller Foundation', in Katherine Angel, Edgar Jones and Michael Neve (eds.), *European Psychiatry on the Eve of War: Aubrey Lewis, the Maudsley Hospital and the Rockfeller Foundation in the 1930s* (London: Wellcome Trust Centre for the History of Medicine at University College London, Medical History Supplement 22), pp. 39–56.

p. 22 *'It's author, Ewen …'* The programme was not a success. See E. Cameron, J. G. Lohrenz and K. A. Handcock, 'The Depatterning Treatment of Schizophrenia', *Comprehensive Psychiatry* 3(2) (April 1962), 65–76.

p. 23 *'funding from the CIA …'* Anne Collins, *In the Sleep Room: The Story of CIA Brainwashing Experiments in Canada* (Toronto: Key Porter Books, [1988] 1998) pp. 39, 42–3, 133.

p. 25 *'loss of memories …'* I. Janis, 'Psychologic Effects of Electric-convulsive Treatments', *Journal of Nervous and Mental Diseases* 3(6) (1950), 469–89.

p. 28 *'Lucy Tallon, a …'* Lucy Tallon, 'What is having ECT like?', *Guardian G2*, 14 May 2012.

p. 28 *'quoting Carrie Fisher …'* Carrie Fisher, *Shockaholic* (New York: Simon & Schuster, 2011).

p. 28 *'all physicians, yourselves …'* Sigmund Freud, 1904, published in *Collected Papers Vol. 1* (London: Hogarth Press, 1953).

3. Eye: A Renaissance of Vision

p. 32 *'As when a man …'* Empedocles, 'On Nature', Fragment 43, in *The Fragments of Empedocles*, translated by William Ellery Leonard (Chicago: Open Court Publishing Company, 1908).

p. 36 *'Ophtalmologists struggle to …'* J. García-Guerrero, J. Valdez-García and J. L. González-Treviño, 'La Oftalmología en la Obra Poética de Jorge Luis Borges', *Arch Soc Esp Oftalmol* 84 (2009), 411–14.

p. 36 *'scharlach, scarlet, ecarlata …'* Jorge Luis Borges, 'Blindness', in *Seven Nights* (New York: New Directions, 1984).

p. 40 *'a book on …'* John Berger and Selçuk Demirel, *Cataract* (London: Notting Hill Editions, 2011).

p. 40 *'The image of …'* John Berger, 'Who is an Artist?', in *Permanent Red: Essays in Seeing* (London: Methuen, 1960), p. 20.

p. 40 *'Shelf of a field …'* John Berger, 'Field', in *About Looking* (London: Writers and Readers Cooperative, 1980) p. 192.

4. Face: Beautiful Palsy

p. 46 *'the joint of …'* 'La giuntura delli ossi obbediscie al nervo, e'l nervo al muscolo, e'l muscolo alla corda, e la corda al senso comune, e'l senso comune è sedia dell'anima', Leonardo W. 19010r, quoted after Richter Literary Works §838.

p. 47 *'And you, man …'* From folio 2, recto, of the anatomical drawings in the Royal Collection.

p. 50 *'One of the leaves …'* Martin Clayton and Ron Philo, *Leonardo da Vinci: The Mechanics of Man* (London: Royal Collection Trust, 2013).

p. 53 *'when forty winters …'* Sonnet 2.

p. 53 *'creased like a …'* Iain Sinclair, *Landor's Tower* (London: Granta, 2002), p. 120.

p. 54 *'who taught the boy …'* Charles Bell, *Letters of Sir Charles Bell: selected from his correspondence with his brother, George Joseph Bell* (London: John Murray, 1870).

p. 54 *'a new system …'* Charles Bell, *A System of Dissections* (Edinburgh: Mundell & Son, 1798). Vesalius' masterpiece was *De humani corporis fabrica* (Of the fabric of the human body) (1543).

p. 55 *'the sketches he …'* M. K. H. Crumplin and P. Starling, *A Surgical Artist at War: the Paintings and Sketches of Sir Charles Bell 1809–1815* (Edinburgh: Royal College of Surgeons of Edinburgh, 2005).

p. 56 *'later published as …'* Charles Bell, *Essays on the anatomy of the expression in painting* (London: John Murray, 1806). Later published as *Essays on the anatomy and philosophy of expression as connected with the fine arts* (1844).

p. 57 *'Bell may with …'* Charles Darwin, *The Expression of the Emotions in Man and Animals* (London: John Murray, 1872).

p. 57 *'your painting will prove …'* Author's translation from Leonardo da Vinci's *Trattato della pittura*, loosely adapted from the English translation of 1721 by John Senex.

p. 59 *'borne by psychologists …'* James D. Laird, 'Self-attribution of emotion: The effects of expressive behavior on the quality of emotional experience', *Journal of Personality and Social Psychology* 29(4) (April 1974), 475–86.

5. Inner Ear: Voodoo & Vertigo

p. 66 *'published in a journal …'* J. M. Epley, 'The canalith repositioning procedure: for treatment of benign paroxysmal positional vertigo', *Otolaryngol – Head and Neck Surgery* 107(3) (September 1992), 399–404.

7. Heart: On Seagull Murmurs, Ebb & Flow

p. 90 *'General anaesthesia …'* Robin Robertson, 'The Halving', in *Hill of Doors* (London: Picador, 2013).

8. Breast: Two Views on Healing

p. 96 *'the healing that …'* Brigid Collins, October 2014, personal communication.

p. 96 *'the exhibition …'* The exhibition *Frissure* took place at the Scottish Poetry Library, November 2013. A book of images and text is published by Polygon (Edinburgh: 2013).

9. Shoulder: Arms & Armour

p. 106 *'Hector jumped down …'* quote from *The Iliad* adapted by the author from Samuel Butler's 1898 translation.

p. 107 *'there are some medically …'* E. Apostolakis et al., 'The reported thoracic injuries in Homer's *Iliad*', *Journal of Cardiothoracic Surgery* 5 (2010), 114. See also A. R. Thompson, 'Homer as a surgical anatomist', *Proceedings of the Royal Society of Medicine* 45 (1952), 765–7.

p. 110 *'A historian of …'* P. B. Adamson, 'A Comparison of Ancient and Modern Weapons in the Effectiveness of Producing Battle Casualties', *Journal of the Royal Army Medical Corps* 123 (1977) 93–103.

10. Wrist & Hand: Punched, Cut & Crucified

p. 120 *'Teenagers admit to …'* Edward Hagen, Peter Watson and Paul Hammerstein, 'Gestures of despair and hope: A view on deliberate self-harm from economics and evolutionary biology', 2008, philpapers.org.

p. 121 *'As the blood flows …'* J. Harris, 'Self-harm: Cutting the bad out of me', *Qualitative Health Research* 10 (2000), 164–73.

p. 121 *'a strategy of withdrawal …'* F. X. Hezel, 'Cultural patterns in Truckese suicide', *Ethnology* 23 (1984), 193–206.

p. 122 *'communication of emotional …'* A. Ivanoff, M. Brown and M. Linehan, 'Dialectical behavior therapy for impulsive self-injurious behaviors', in D. Simeon & E. Hollander (eds.), *Self-injurious behaviors: Assessment and treatment* (Washington DC: American Psychiatric Press, 2001).

p. 127 *'Barbet found that …'* Pierre Barbet, *Les Cinq Plaies du Christ,* (Paris: Procure du carmel de l'action de grâces, 1937).

p. 127 *'the professor of anatomy ...'* Nicu Haas, 'Anthropological Observations on the Skeletal Remains from Giv'at ha-Mivtar', *Israel Exploration Journal* 20 (1970), 38–59.

p. 128 *'came to different ...'* Joseph Zias and Eliezer Sekeles, 'The Crucified Man from Giv'at ha-Mivtar: A Reappraisal', *Israel Exploration Journal* 35(1) (1985), 22–7.

p. 128 *'I've read them ...'* C. J. Simpson, 'The stigmata: pathology or miracle?', *British Medical Journal* 289 (1984), 1,746–8.

11. Kidney: The Ultimate Gift

p. 133 *'De Zerbis was ...'* Richard Eimas (ed.), *Heirs of Hippocrates* (Iowa City: University of Iowa Press, 1990): entry no. 137, GABRIELE DE ZERBIS (1445–1505), *Gerentocomia* [1489].

12. Liver: A Fairy-Tale Ending

p. 150 *'if he were opened ...'* Speech by Sir Toby Belch, *Twelfth Night*, Act III, Scene 2.

p. 150 *'For the king of Babylon ...'* Ezekiel 21:21.

p. 153 *'sweet, biddable girls ...'* Marina Warner, 'How fairy tales grew up', *Guardian Review* (13 December 2014).

13. Large Bowel & Rectum: A Magnificent Work of Art

p. 161 *'According to the psychology ...'* Paul J. Silvia, 'Looking past pleasure: Anger, confusion, disgust, pride, surprise, and other unusual aesthetic emotions', *Psychology of Aesthetics, Creativity, and the Arts* 3(1) (February 2009), 48–51.

14. Genitalia: Of Making Babies

p. 167 *'would have no desire ...'* Mrs Jane Sharp, *The Midwives Book* (1671) is referenced in Thomas Laqueur, 'Orgasm, Generation, and the Politics of Reproductive Biology', in *The Making of the Modern Body: Sexuality and Society in the Nineteenth Century*, edited by Catherine Gallagher and Thomas Laquer (Berkeley: University of California Press, 1992). I owe a great deal to Professor Laqueur's paper for many of the ideas explored in this chapter.

p. 167 *'Of the commingling ...'* Marquis de Sade, *La philosophie dans le boudoir* (1795).

p. 167 *'so that she can …'* Rachel Maines, *The Technology of Orgasm: 'Hysteria,' the Vibrator, and Women's Sexual Satisfaction* (Baltimore: Johns Hopkins University Press, 1999).

p. 172 *'the area around the …'* Giovanni Luca Gravina et al., 'Measurement of the Thickness of the Urethrovaginal Space in Women with or without Vaginal Orgasm', *The Journal of Sexual Medicine* 5(3) (March 2008), 610–18.

p. 172 *'Like Ernst Grafenberg …'* Ernst Gräfenberg, 'The Role of the Urethra in Female Orgasm', *International Journal of Sexology* 3(3) (February 1950), 145–8.

p. 173 *'so the poetic …'* Arthur Aikin (ed.), *The Annual Review and History of Literature for 1805, Volume IV* (London, 1806).

p. 174 *'Women's psychology is …'* Carl Jung, 'Women in Europe', in *Collected Works of C. G. Jung, Volume 10: Civilization in Transition*, edited and translated by Gerhard Adler and R. F. C. Hull (Princeton University Press, 1970), p. 123.

p. 176 *'do not fulfil …'* Avicenna's *Canon* 3:20:1:44.

p. 177 *'the man is quicke …'* John Sadler, *The Sicke Woman's Private Looking Glass* (London, 1636), p. 108.

p. 177 *'a paper appeared …' The Lancet*, 28 January 1843, p. 644.

p. 177 *'Marie Stopes bestselling …'* Marie Stopes, *Married Love* (London: A. C. Fifield, 1919).

16. Afterbirth: Eat It, Burn It, Bury it under a Tree

p. 190 *'One might recall …'* Herodotus, *Histories* 3:38, in Aubrey de Selincourt's Penguin Classics Translation (Harmondsworth, 1954).

p. 191 *'From Morocco to Moravia …'* E. Croft Long, 'The Placenta in Lore and Legend', *Bulletin of the Medical Library Association* 51(2) (1963), 233–41.

p. 192 *'I was born with …'* Charles Dickens, *David Copperfield* (London: Bradbury & Evans, 1850).

p. 192 *'with earth piled …'* James Frazer, *The Golden Bough*, third edition (Cambridge University Press, 2012), p. 194.

p. 192 *'The Russians traditionally …'* Barbara Evans Clements, Barbara Alpern Engel and Christine Worobec (eds.), *Russia's Women: Accommodation, Resistance,*

Transformation (Berkeley and Los Angeles: University of California Press, 1991), p. 53.

p. 194 *'read an essay …'* Seamus Heaney, 'Mossbawn', in *Finders Keepers: Selected prose 1971–2001* (London: Faber & Faber, 2003).

17. Hip: Jacob & the Angel

p. 199 *'He had studied …'* Italo Svevo, *La Coscienza di Zeno* (Milan: Einaudi, 1976), p. 109 (author's translation).

p. 202 *'If someone over the …'* J. A. Grisso et al., 'Risk Factors for falls as a cause of hip fracture in women', *The New England Journal of Medicine* (9 May 1991), 1,326–31.

p. 202 *'Around 40 percent …'* Figures from Atul Gawande, *Being Mortal: Medicine and What Matters in the End* (London: Profile, 2014).

p. 202 *'Between five and eight …'* P. Haentjens et al., 'Meta-analysis: Excess Mortality After Hip Fracture Among Older Women and Men', *Annals of Internal Medicine* 152 (2010), 380–90.

p. 203 *'His name Yaakov …'* My reading of Jacob's story has been informed by Geoffrey H. Hartman, 'The Struggle for the Text', from Geoffrey H. Hartman and Sanford Budick (eds), *Midrash and Literature* (London: Yale University Press, 1986), pp. 3–18.

p. 203 *'Some commentators have …'* Roland Barthes, 'The Struggle with the Angel', in *Image, Music, Text*, translated by Stephen Heath (Glasgow: Fontana Press, 1977). See also the theories of Vladimir Propp on the universal problems of folk tales.

18. Feet & Toes: Footsteps in the Basement

p. 219 *'A halo of wonder …'* Virginia Woolf, 'The Elizabethan Lumber Room', in *The Common Reader* (London: The Hogarth Press, 1925).

List of images

p. 62 Interior of right osseous labyrinth (figure 921 in *Gray's Anatomy*, 1918 edition).

p. 73 'Bronchial tube with its bronchules', from *Popular Science Monthly* (1881).

p. 77 Chest X-ray (this one suggests a developing right upper lobe pneumonia rather than subcarinal lymphadenopathy). From Public Health Library (no. 5802), by Dr Thomas Hooten (1978).

p. 79 Laryngoscopic view of interior of larynx (figure 956 in *Gray's Anatomy*, 1918 edition).

p. 83 Section of the heart showing the ventricular septum (figure 498 in *Gray's Anatomy*, 1918 edition).

p. 86 Aorta laid open to show the semilunar valves (figure 497 in *Gray's Anatomy*, 1918 edition).

p. 94 Dissection of the lower half of the mamma during the period of lactation (luschka) (figure 1172 in *Gray's Anatomy*, 1918 edition).

p. 97 'Cancer of the breast, field operation, just before the final cut'. Wellcome Image Collection.

p. 98 *Dog Rose* by Brigid Collins, reproduced by kind permission of the artist.

p. 99 *Kist* by Brigid Collins, reproduced by kind permission of the artist.

p. 100 *In September* by Brigid Collins, reproduced by kind permission of the artist.

p. 106 A fractured right collarbone, reproduced here thanks to Sam Woods of the Fruitmarket Gallery, Edinburgh.

p. 108 The right brachial plexus with its short branches, viewed from the front (figure 808 in *Gray's Anatomy*, 1918 edition).

p. 117 *Quatre Mains,* ink drawing by Yves Berger, reproduced from the book *Caring* (Galleria Antonia Jannone, October 2014).

p. 120 Detail from a copy of Rembrandt's *The Anatomy Lesson of Dr Nicolaes Tulp,* which hangs at the entrance to the Anatomy Museum, University of Edinburgh.

p. 127 Image reproduced from Pierre Barbet, *Les Cinq Plaies du Christ* (Paris: Procure du carmel de l'action de graces, 1937) p. 63.

p. 133 Diagrams from the work of Vesalius, demonstrating orthodox belief.

p. 135 Microscopic view of a glomerulus, the filtering unit of the kidney (figure 1130 in *Gray's Anatomy*, 1918 edition).

p. 141 The illustration on the kidney donor gift circle was produced by the author.

p. 144 From Alec Finlay, *Taigh: A Wilding Garden* (Edinburgh: Morning Star Publications, 2014).

p. 145 Ibid.

p. 147 Grid of biochemistry results, author's collection.

p. 152 'Snow White', illustrated by Walter Crane (1882).

p. 159 Barium enema, courtesy of Diagnostic Image Centers, Kansas City.

p. 162 French poster of glass bottles, varying shapes. Public domain.

p. 168 'Demonstration using the vibrator, 1891' from 'A description of the vibrator and directions for use'. Wellcome Image Collection.

p. 188 Foetus attached to umbilical cord and placenta. Wellcome Image Collection.

p. 193 Human placenta. Wellcome Image Collection.

p. 200 'The bones and muscles of the hip and thigh', drawing, 1841.Wellcome Image Collection.

p. 204 'X-ray of Right Total Hip Replacement', from the US National Institute of Health.

p. 211 *Romería* poster from 2011, courtesy of Chemi Marquez.

p. 213 Skeleton of the foot (medical aspect) (figure 290 in *Gray's Anatomy*, 1918 edition).

p. 221 Photograph by Edwin Aldrin, Apollo 11, courtesy of NASA.

Index

C

WELLCOME COLLECTION is the free visitor destination for the incurably curious. It explores the connections between medicine, life and art in the past, present and future. Wellcome Collection is part of the Wellcome Trust, a global charitable foundation dedicated to improving health by supporting bright minds in science, the humanities and social sciences, and public engagement.